Improve and Accessories for your Lathe

by

J A Radford

TEE Publishing

ERRATA:
As a result of a computer malfunction, the drawing on page 58 overlaid an entire page of text which was then omitted. Supplementary page 58a now overlays page 59.

ISBN No. 1 85761 105 5

Originally published as separate articles in *Model Engineer*, 1967-1971

Contents

CHAPTER 1

A Milling Attachment for the Lathe

I bought the Myford milling fixture with the universal vertical slide for my Super 7 and set it up according to recommendations. But when the vertical slide was swung completely down and parallel to the indexing head and at an angle to the cross-slide, it soon became apparent that this was not very satisfactory. In setting up with the slide in a vertical position, there was insufficient room to mill a gear of any size and also the amount of cross-slide traverse left in front of the milling head was very small.

It seemed to me to be necessary to bring the base of the vertical slide much higher and right back towards the operator to the end of the cross-slide and also to bring the base beyond the right-hand side of the cross-slide to reduce the overhang. As there is only one tee slot in the end of the Super 7 cross-slide and then the large hole to take the topslide, the means for securing the vertical slide to the cross-slide seemed to me to be inadequate.

Accordingly, I designed and made two bases to fit the cross-slide and also both the universal and the fixed vertical slides, bringing the bases of the vertical slides to the best positions and providing very sound and strong means of support. At the same time I discarded the screwed mandrel provided in the milling attachment and made a collet type which protruded only half an inch from the end of the bearing and I made several collets and mandrels to fit. Also I tapped the end of the bar support for a $5/16$" Whit. screw and made the bar support bracket shown in my drawing and in one of the photographs.

This arrangement, especially when using the fixed vertical slide, enabled much useful gear cutting and some keyways in small shafts to be done. However, although I could now mill the full length of the cross-

A view of the patterns.

Pulleys 3/8" wide
1/4", 1", 7/8" O.D.
Drill & top large
pulley for 2 BA
grub screw

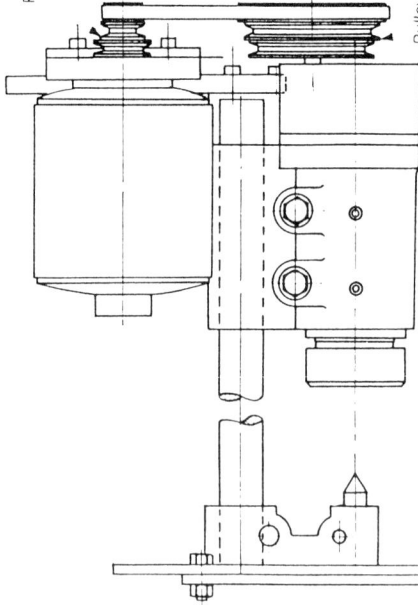

Pulleys 3/8" wide
3/4", 3", 2 3/4" O.D.
Material C.I.

G.A. MOTORISED MILLING ATTACHMENT FOR THE LATHE

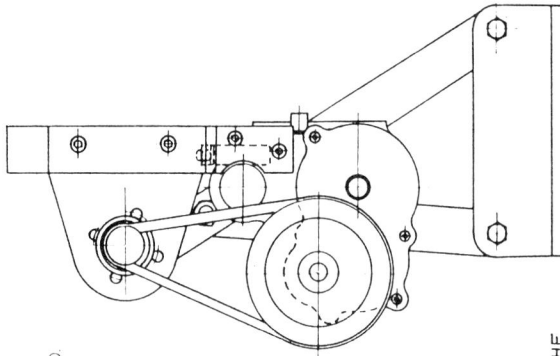

1005

MODEL ENGINEER   20 October 1967

slide travel, which was 6", and accurately determine the proper depth of the cutter in a vertical direction, I was still very limited in the lengths which it was possible to mill and also, as the cutter was underneath the work, it was impossible to see what was going on; a mistake in indexing did not become apparent until the work had progressed some distance around the blank.

The ideal milling attachment, it seemed to me, was one that had a cutter which moved with the saddle and cross-slide, so that a blank could be turned, milled in place, finished and parted off, saving all the trouble of making mandrels, etc., and where a long shaft could be turned, all keyways milled in place, whether of the slotted type or end-milled feather type, before being shifted in the lathe. Such keyways would be true and parallel in both planes. Also such things as dog clutches should be simple to mill, the making of my own milling cutters should be

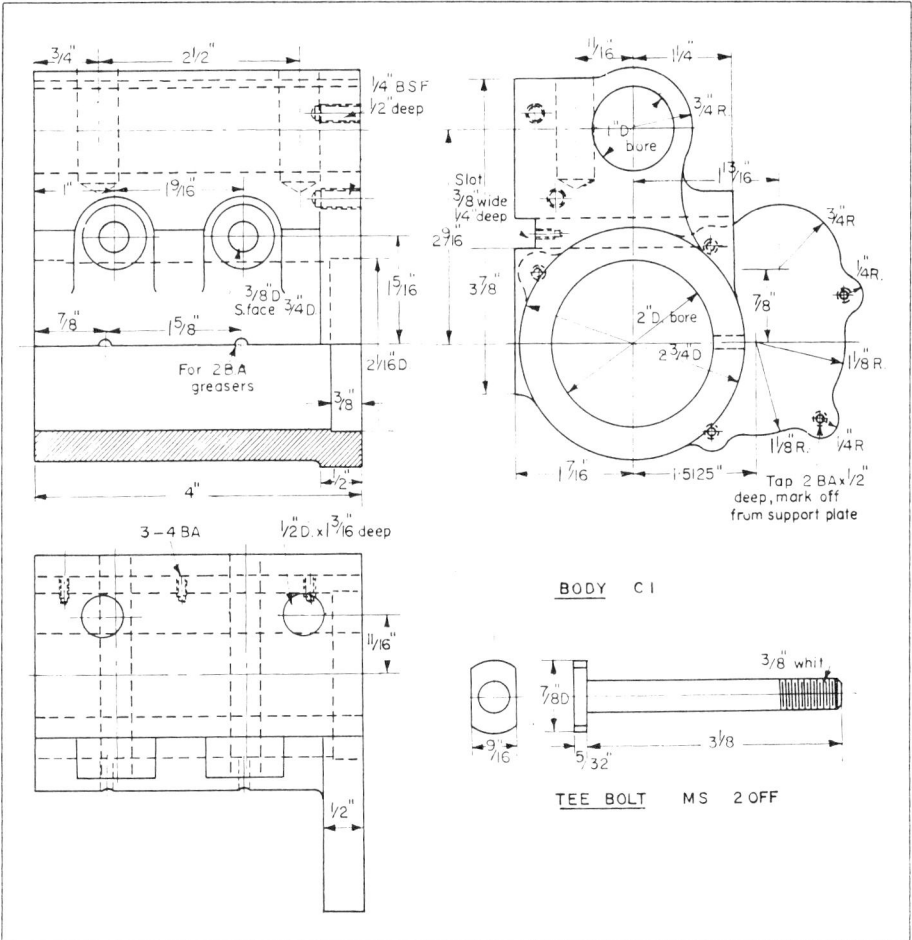

BODY C I

TEE BOLT   MS 2 OFF

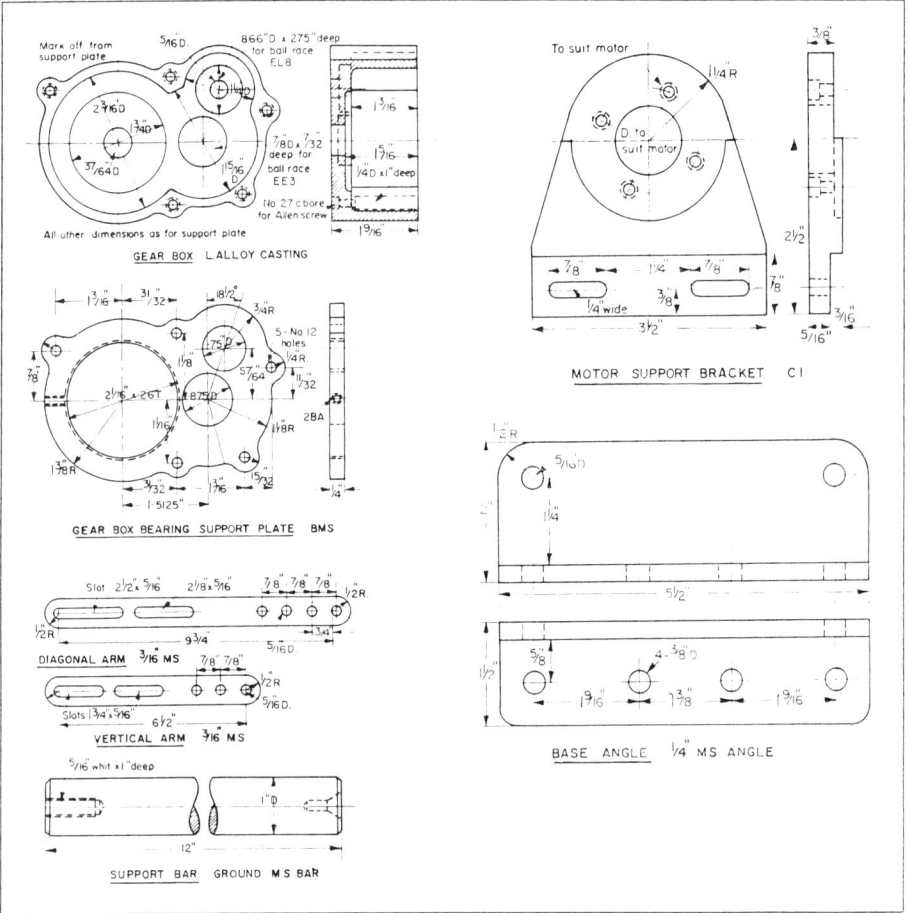

GEAR BOX LALLOY CASTING

GEAR BOX BEARING SUPPORT PLATE BMS

DIAGONAL ARM 3/16 MS

VERTICAL ARM 3/16 MS

SUPPORT BAR GROUND MS BAR

MOTOR SUPPORT BRACKET CI

BASE ANGLE 1/4 MS ANGLE

possible and also it seemed to me that if a long casting could be set up parallel to the rear edge of the bed and in a vertical position, it should be possible to mill such things as machine slides to the full length of the saddle travel and perhaps with automatic feed.

I considered many different types of drive for a long time. Overhead shafts, overhead motors on swinging brackets, flexible shaft drive, even air drive, but none seemed to be very satisfactory. A motor drive seemed to be the only solution, but how could a small enough motor that could be accommodated on the lathe provide sufficient power to do a useful amount of work?

Investigating my 5/16" Desoutter 860 rev. electric drill with a 9-1 speed reducer, I thought I had a sufficient amount of torque to do a great deal of milling. This was too awkward to set up as a drive, but I investigated several series-wound motors. I found one with a current consumption of

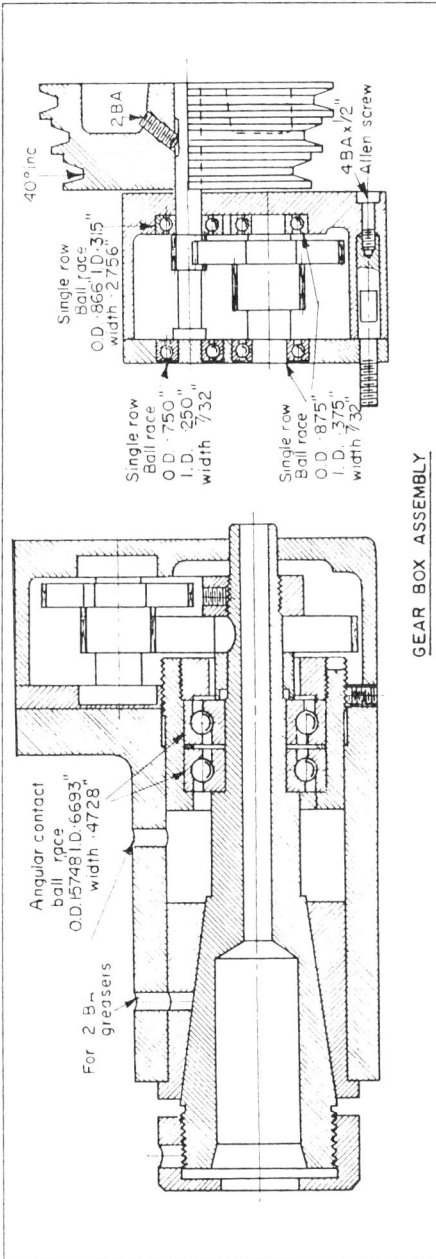

GEAR BOX ASSEMBLY

Labels in diagram:
40° inc
2BA
4BA x ½" Allen screw
Single row Ball race O.D. .866" I.D. .315" width .2756"
Single row Ball race O.D. .750" I.D. .250" width 7/32
Single row Ball race O.D. .875" I.D. .375" width 7/32
Angular contact ball race O.D. .15748 I.D. .6693" width .4728"
For 2 B greasers

1 amp. (the drill takes 0.75 amp.) so with 33⅓ per cent more power and at a cutter speed of around 100 r.p.m. I felt that I had sufficient torque, provided that the speed reduction from 4,500 to 100 r.p.m. could be done efficiently. Worm gear seemed the obvious solution but I ruled it out because of friction losses. Spur gearing running in ball races (as in the electric drill) with an initial reduction of approx. 3-1 through three step cone pulleys and a ¼" vee belt, leaving 15-1 for the spur gears, seemed ideal.

I had sufficient confidence to go ahead and make patterns and complete the whole attachment using the existing baseplate and vertical slide and bar support bracket and I have been fully justified. I had not the slightest trouble milling the 16 D.P. drive gears for the Allchin traction engine, making the bevel gears for the differential, making 2½" dia. x ⅝" facing milling cutters in H.S. steel and many other cutters, and the final justification was milling the slides complete for a 'Bormilathe' type of attachment illustrated in one of the photographs. I have since made a quartering table 6¼" square to fit the cross-slide, so I now have a useful horizontal borer, all made possible by this milling attachment.

I will now describe the construction of the various parts, knowing that those readers who have the ability and desire to construct this milling attachment will find that it broadens the scope of their work enormously.

The patterns illustrated in one of the photographs show the two baseplates with the core box for the fixed vertical slide base, the back centre support arm, the main body casting, the gearbox pattern with the motor support bracket behind, and the pulley pattern behind which is the distance block. The core box shown (the only one required) is made of two pieces of wood ⅞" thick. The shape is the shape of the core print on the pattern, plus the desired internal shape. The moulder merely lays the two pieces clamped together on a flat piece of iron, fills with sand, wipes off the top with a straight edge and puts the sand, after opening and removing the core box, in the oven to bake.

Allow $1/16$" on all surfaces for machining ($1/8$" radius on round cored holes) and finish the patterns with several coats of a thin mixture of french polish (bought in tins) and lamp black sanded between each coat, with a final coat of french polish only and the core prints with a thin mixture of red lead and French polish. The gearbox is made up on the bread and butter principle, the end cover going on after the internal faces are finished – taper the inside well. You will notice that several lugs are made on the end cover before being glued and dowelled

in place. These are for holding in various positions for machining as will be described later. The fillets are made by running beeswax into the corners with a small electric soldering iron, after which the fillets are scraped to shape with a scraper made from a piece of broken hacksaw blade. Sand smoothly all patterns with very fine glass paper before varnishing.

The baseplates are faced in the lathe on the 9" faceplate. I marked out and drilled the tee bolt holes before turning, as the holes are useful for holding down. The top was turned first, it will clear the bed all right, and was then drilled and tapped $3/8$" Whit. in the

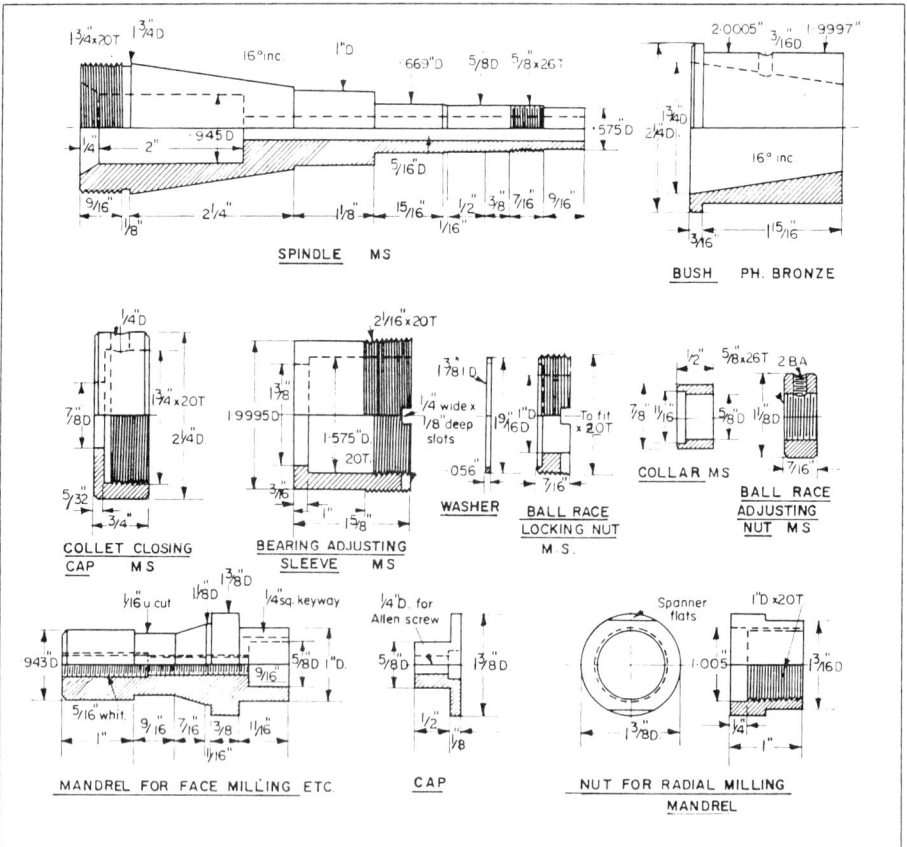

SPINDLE MS

BUSH PH. BRONZE

COLLET CLOSING CAP MS

BEARING ADJUSTING SLEEVE MS

WASHER

BALL RACE LOCKING NUT M.S.

COLLAR MS

BALL RACE ADJUSTING NUT MS

MANDREL FOR FACE MILLING ETC.

CAP

NUT FOR RADIAL MILLING MANDREL

The 'Bormilathe' type of elevating heads completed.

correct places and this face set against the faceplate for turning the base, this also will clear the bed. A counterweight which also acts as a driving lug was used. A third hole you will notice was tapped for a third screw for the fixed steady and two dowel holes in each base. The final adjustment, to bring the vertical slide dead parallel to the faceplate after the surfaces had been hand scraped to surface plate, was accomplished by some filing and fitting of the $5/16$" grooves into which are set the short pieces of $5/16$" key steel. These bear against the right-hand edge of the cross-slide and ensure that the vertical slide is always at right angles to the bed. The undersides of the bases are recessed to make for less machining and scraping and the tee bolt holes are rather deeply spot faced. The universal slide base has a $7/16$" drilled hole and the bolt, with 1" dia. head, is held in place and stopped from turning with a 2BA grub screw drilled and

tapped in place after the $7/16$" bolt has been pulled up tightly by means of a collar and nut. No core is required for this base but the webs should be tapered downwards as shown.

When drilling the hole in the fixed vertical slide for the third holding down bolt, make it $7/16$" instead of $3/8$" (but not in the base) as this slide can then be used on the topslide of the lathe for such things as bevel gears, facing cutters, etc., as it has considerably less overhang than the universal type. Now the main body casting can be machined.

The base and keyway should be machined in one setting. I carefully marked out for the two holding down bolts, working from the outer diameters of the main cored hole, not from the actual cored hole, and instead of drilling these holes $3/8$", I drilled $5/16$" and tapped $3/8$" and bolted through with the distance pieces on my 6" hand shaper with

11

additional clamps in the edges of the main cored hole. The face was planed and the $3/8$" keyway cut through the centre of these two holes to a depth of $1/4$". The holes were then drilled $3/8$" and spot faced.

The boring was done in the lathe by bolting on the cross-slide. Two pieces of $5/8$" key steel were required for packing to centre height and two pieces of $3/4$" × $3/8$" bright steel were tightly fitted into the keyway at each end and into one of the cross-slide tee slots to align the casting for boring true to the keyway. Two long tee bolts were made and an additional clamp or two was used. The casting was set up so that both holes could be bored at one setting at $2 9/16$" centres using the cross-slide index. The large hole was bored 2" using a boring bar with a 2" long end turned to 1" dia.; the centre part was $1 1/2$" dia. with the $1/4$" dia. H.S. tool in the centre of the length of $9 1/2$" and the end was about 1" long, turned to a large radius down to 0.590" to fit a ball race held in a split steel collar clamped on the end of the tailstock barrel.

I prefer this method for all boring bars and milling arbors in preference to using a centre. The fitting is similar to that shown as the 'alternate arbor support', but has a split hole to fit the tailstock barrel clamped with

an Allen screw instead of the straight shank shown.

The end of the 2" hole is counterbored just over $2 1/16$" dia. for $3/8$" long to clear the thread in the screwed sleeve. By the way, set up this end with the integral cast plate to finish to $1/2$" thick, next to the chuck so that this face can be finished at the same setting as the boring. The small hole was bored to a good sliding fit on 1" ground m.s. bar. I used a $5/8$" dia. bar first and then a $7/8$" to finish. Use H.S. round steel for cutters, not silver steel.

The integral face is milled with a large diameter facing cutter if you have one, or with a flycutter, which is almost as good. The total length of the casting should finish to 4".

There is one detail that I had forgotten which should be done before boring: the two bar clamps should be made and fitted into $1/2$" reamed holes before the hole for the bar support is bored. These two clamps are held in place temporarily with a 4BA grub screw until the boring is done when these screws are returned to stock. These two clamps are $25/32$" long in the sleeve part and clamped tightly together for boring and afterwards $1/16$" is turned off the sleeves to allow for clamping up. This method of machining

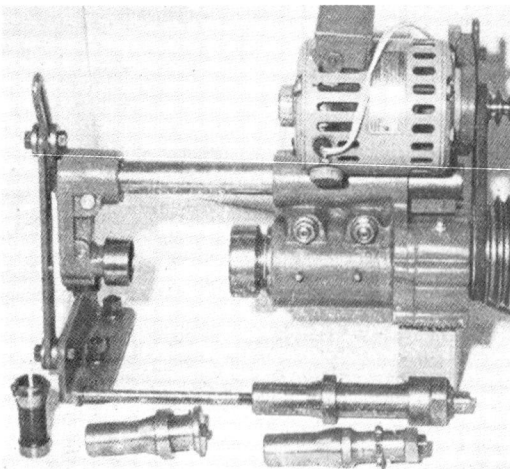

A general view of the milling attachment and some accessories.

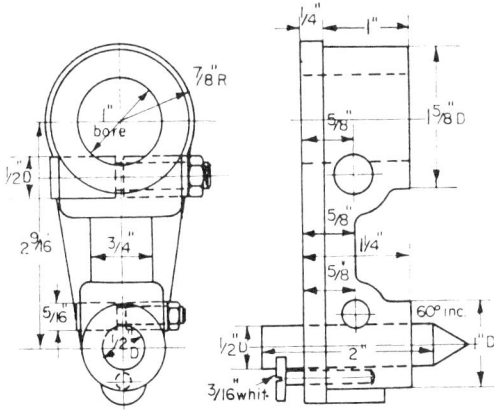

BACK CENTRE SUPPORT ARM   CI

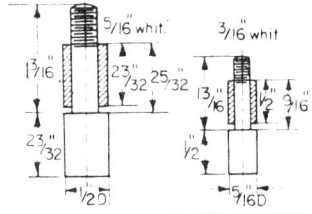

SUPPORT BAR CLAMP BRONZE

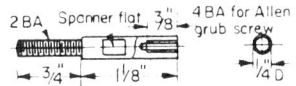

GEAR BOX BOLT   MS 5 OFF

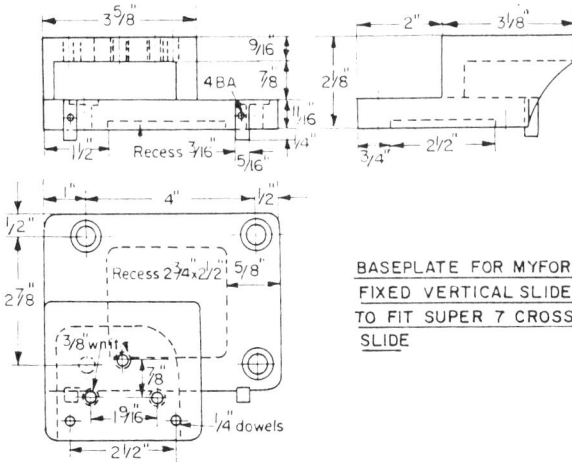

BASEPLATE FOR MYFORD
FIXED VERTICAL SLIDE
TO FIT SUPER 7 CROSS
SLIDE

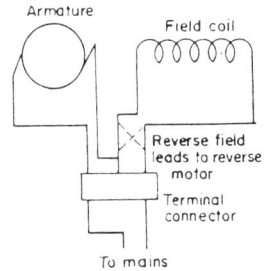

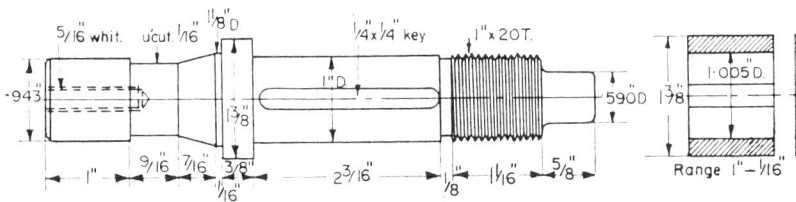

MANDREL FOR RADIAL MILLING, GEAR CUTTING ETC.

COLLAR

BASEPLATE FOR MYFORD
UNIVERSAL VERTICAL
SLIDE. TO FIT SUPER 7
CROSS SLIDE

Once the topslide is set over for the taper of 8 deg. a side, it should not be touched until the spindle is finished. The outside and inside and face of the bush are finished completely before parting off from the chuck.

The spindle is now made from a piece of $1^7/8$" or 2" mild steel or perhaps something a little better quality, but use the free cutting type if obtainable. Do not bore the hole but rough down all over to, say, $1/16$" over each size and then finish accurately to size. The ball races 7203B should be a fairly tight push fit. The keyway for the driving gear is made with a $1/2$" × $1/8$" woodruffe cutter. The key is made from H.T. steel to stand proud slightly under $1/8$". Do not cut the 20 t.p.i. on the $1^3/4$" dia. end until the spindle is bored, as this end will have to ride in the fixed steady for boring; the $5/8$" by 26 t.p.i. thread should, however, be screw-cut.

makes a perfect fit around the support bar. A piece of cored phosphor-bronze or lead bronze 3" long to finish to $2^1/4$" dia. and with a 1" cored hole should be obtained.

The reason for making the bush before the spindle is that the bush can be tried on to the spindle, but the spindle cannot be passed through the bush in the lathe. The outside should be tapered $3/4$ thou. and the small end should just screw in the end of the bored casting about $1/2$"; as this will be at the end next to the chuck a plug gauge should be made, if you are not confident that you can measure this accurately enough. I was perhaps lucky with mine.

The boring and drilling is done in the fixed steady with the small end in the four-jaw chuck; this end should be set true to the dial gauge. Before the boring is done, however, the spindle should be polished to a high finish and should fit the taper bush perfectly leaving $1/16$" of spindle taper showing at each end of the bush. All the lengths are most important and care should be taken over them. After boring and taper boring for the collets, which are standard 24mm collets obtainable from any Myford agent, the thread for the collet closing cap can he screw-cut after a true plug or short milling arbor has been made to fit.

The bronze bush is driven into the body using the rest of the bush truly faced each side and a piece of hardwood. No other fixing is necessary. The oil hole is drilled and tapped and a cross oil groove cut through this hole. The bush may have closed up slightly and if

INTERMEDIATE PINION   INTERMEDIATE GEAR   SPINDLE GEAR

so should be hand scraped back to the proper fit and distance.

The bearing adjusting sleeve is made a good push fit in the body and don't forget that the external thread is $2^1/_{16}$" × 20 t.p.i. The internal bore is made a push fit only on the bearings and then the internal thread of 20 t.p.i. is cut $9/_{16}$" long. Any burr left by the thread should be scraped out to leave the bearings free to enter. Four slots $1/_4$" wide by $1/_8$" deep are cut in the end; this is a case where the milling attachment would be useful, but I had to cut them by hand. The washer in the centre is to keep the bearings apart so they can be adjusted. Note the way the bearings are assembled in the sleeve. The ball race locking nut is made and once the ball races and washer are in place this nut can be locked up tightly and requires no further adjustment. It would be well to explain the adjustment for both the ball races and the taper bearing.

The ball race adjusting nut bears against the spindle gear and through the collar closes the inner parts of the ball races together, thus taking up any wear in these bearings. The adjustment is made by first screwing the bearing adjusting sleeve complete with its assembled ball races and with the spindle in place, well clear of the tapered bronze bush. The ball races are adjusted until they can just be felt. The lock nut is then slackened back about a quarter turn and the 2BA grub screw with its brass pad locked up. Then by turning the bearing adjusting sleeve outwards to bring the spindle into contact with its taper bush, this bearing can be adjusted. The adjustment should leave the spindle perfectly free but without shake. The end thrust either way against the spindle is taken up by the ball races and

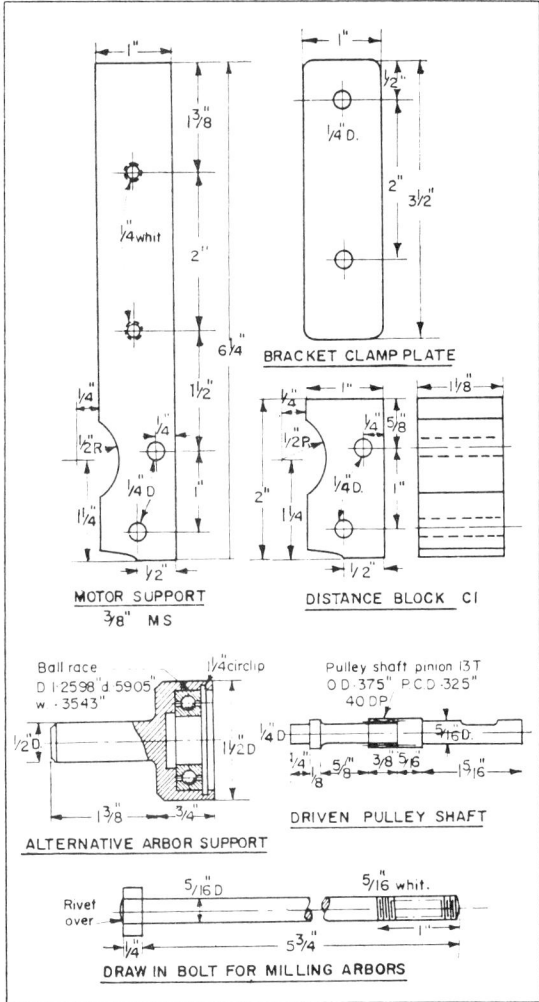

BRACKET CLAMP PLATE

MOTOR SUPPORT
$3/_8$" MS

DISTANCE BLOCK CI

ALTERNATIVE ARBOR SUPPORT

DRIVEN PULLEY SHAFT

DRAW IN BOLT FOR MILLING ARBORS

cannot jam the spindle into its taper bearing.

The principle is the same as used in the Myford Super 7 and indeed the bush is practically the same size although the ball races are smaller. It makes an excellent bearing. There is a 2BA Allen grub screw with brass pad in the steel bearing support plate to prevent the bearing adjusting sleeve

Showing the method of milling the slides, the pattern has been sdt up in lieu of the actual casting.

from turning once the adjustments have been made.

The gears and gear spindles are all made from 3 per cent nickel case-hardening steel, case-hardened after machining. The reason for the seemingly odd numbers of teeth is that with milling the load is periodic, and if this period coincides with the number of teeth in a gear, wear will take place chiefly on that gear; if it is a multiple of its mating gear, wear will be transmitted to this gear also. With an odd number of teeth, every tooth mates with every tooth in its mating gear and so any wear is evenly spread throughout the whole gear. The teeth are all 40 D.P. 14$\frac{1}{2}$ deg. pressure angle. A No. 8 cutter is required for the 13-tooth pulley shaft pinion, a No. 4 cutter for the intermediate shaft pinion which is 31 T. The intermediate gear is 66 T. and requires a No. 2 cutter and the same cutter does for the spindle gear which is 90 T. If you cannot beg or borrow these cutters and they prove rather expensive to buy, it may pay to have the teeth cut on your blanks rather than attempt to use home-made cutters, which are seldom very satisfactory for this type of work.

Previously I bought just these three cutters for making an electric grandfather clock, so I was lucky. The gears were made with the Myford indexing head set up on the baseplate and using the bar support bracket as described. The intermediate gear 66 T. is made a tight press fit on its spindle, the keyway in this gear is $\frac{1}{8}$" wide and the key is of the peg type made from a short piece of $\frac{3}{16}$" silver steel rod. The main spindle gear of 90 T., while it should be a good tight fit, must still be capable of movement along the shaft while adjusting the ball race adjusting nut.

The gearbox and its accompanying bearing support plate should now be made. This may seem a somewhat troublesome job, but it is not at all difficult if tackled in the following manner: Make the steel plate first from $\frac{1}{4}$" b.m.s. Mark out the outside shape and drill the five No. 13 holes but do not drill any other holes; cut out the outside and file to shape.

Mark out the centre for the large 2$\frac{1}{16}$" threaded hole and draw a line through for the centre of the intermediate shaft bearing hole. The centre distance between the main spindle centre and the intermediate shaft is 1.5125" and between the intermediate shaft and the pulley shaft centre 0.9375". Now if we make three discs of the right diameters with true centre holes and clamp these on the steel plate with their rims touching we can drill through the centres of the discs through the plate and arrive at three holes in the correct centres through the plate. Make one disc 2.025" dia. for the main hole, one disc 1" dia. for the intermediate hole and one disc 0.875" dia. for the pulley shaft hole. Make all three discs about $\frac{3}{8}$" thick with a true $1/4$" dia. hole. Drill the main hole through the plate first with a centre drill, follow with a $15/64$" drill and finally either drill or ream the hole $1/4$". A short piece of silver steel through the plate and the large disc will locate the disc in place and the 1" disc can now be butted against this with the centre line showing through the hole in the disc clamped truly centre and the $1/4$" hole drilled straight through. Another piece of $1/4$" silver steel will hold this in place while the third 0.875" is butted against the 1" disc and also drilled through.

Now grip the open side of the gearbox casting in the four-jaw chuck and face off the plain side with the lugs truly flat. This side is then bolted against the faceplate using clamp plates on the lugs, and the open side is faced off flat to leave the box exactly $1^9/16$" wide. Next lay the steel plate on the open side of the box, clamp and spot 'the five holes, drill No. 32 and tap 4BA. Obtain five countersunk 4BA screws at least $1/2$" long and screw the steel plate on the box. The reason for using the csk. screw is to line the holes up accurately. Now mount the box with the steel plate outermost on the faceplate and set the largest disc in the proper $1/4$" hole using the silver steel rod,

and true this disc up to the dial gauge. Clamp firmly on three or four lugs when running true, remove the disc and bore and screw-cut $2^1/16$" to fit nicely the thread on the bearing adjusting sleeve. Remove the steel plate but do not touch the box holding clamps and carry on boring the box to $2^7/16$" dia. for a depth of $1^1/8$" then to $1^3/4$" for a further depth of $1/8$" and bore right through the balance 0.580"; make a groove in this last part with a screw-cutting boring tool for a little felt oil ring.

Replace the steel plate on the box and reset with the 1" dia. disc in the intermediate bearing hole running true. Bore this hole right through the plate to 0.875" to a nice push fit on one of the EE3 ball races.

Now remove the gearbox plate and, without touching the box, bore out the box as before to $1^{15}/16$" dia. to a depth of $1^5/32$" and bore to 0.875" for the other EE3 bearing to a depth of slightly more than $7/32$", or just under $1/4$", relieve the bottom face of hole about $1/16$" to clear the inner part of the ball race and the end of the intermediate shaft. Once more replace the steel plate on the box and set true with the dial gauge on the 0.875" disc in the third or pulley shaft hole. The steel plate is now bored to 0.750" to fit

Cutting a pinion using the attachment.

A MILLING ATTACHMENT FOR THE LATHE

How the pinion teeth are indexed.

the EE2 ball race right through, the plate removed as before and the box bored to 1¼" dia. for a depth of 1⁵/₃₂" and then bored to 0.866" to fit the EL8 ball race, depth 0.275"; the inner face is relieved as before and finally bored right through 0.315" to clear the ⁵/₁₆" pulley shaft. A small groove bored into this hole will stop any oil from working along the hole.

All that is now necessary to finish the gearbox is to drill out the bolt holes to ¼" for a depth of 1³/₁₆" only from the open face, after first drilling right through from the other face ⁹/₆₄" and then spot face down from the flat face ⁷/₆₄" to bring the heads of the 4BA Allen screws flush with this face. Drilling and machining of light alloy is easily done to a very good finish and accuracy by using soluble oil and water.

Before drilling the bolt holes, however, it may be as well to set up the steel plate again, and file the box to the same contour as the steel plate to make a good looking job. The five bolts can now be made and, after inserting the bearing adjusting sleeve in the main body, screwing the steel plate on its thread, the holes for these bolts can be spotted through and drilled and tapped 2BA in the flange of the body. Again this flange can be filed to the same contour as the steel plate using 2BA screws to hold in place temporarily. A ¹/₆₄" Hallite or drawing paper

gasket goes between the box and steel plate. Note that this zerk oiler and drain plug should be drilled and also a third zerk in the body as shown, this is to oil the ball races on the spindle. Do not use grease for lubrication but a good grade of light automotive oil is the best. Note that on assembly the left-hand edges of the 90 T. gear and the intermediate pinion should be about flush, leaving ⅛" of spindle adjustment for wear between the 90 T. gear and the 66 T. gear.

The gearbox pulley is cast iron and left as heavy as possible for its flywheel effect. The outside diameters are 3¼", 3" and 2¾". The faces are ⅜" wide. Groove with a ⅛" wide parting tool to ³/₁₆" depth then with a tool ground both sides to 40 deg. included, face left and right to ¼" wide at the top so that the belt stands just a few thou proud. Check the angle of your particular belt when pulled around the pulley as these angles sometimes vary. The motor pulley is made from mild steel and these angles will be slightly less than 40 deg., due to the sharper bend. The gearbox pulley is cast with a lug for holding so that all machining can be done at one setting; that is the lug is made on the flat side as in the photograph. This lug is turned first. The hole is reamed with a machine reamer ⁵/₁₆" and the holding screw which bears on a flat on the shaft is 2BA

18

drilled on an angle to just clear the rim for drilling. It is turned all over, of course. The steel motor pulley is 1¼", 1" and ⅞" and a 2BA thread is put through the largest pulley for a grub screw.

The motor is of the flange mounting type which makes it simple to mount and adjust the belt. The motor mounting flange casting or support bracket is bored, faced and drilled to suit the motor and is an iron casting. It is milled along the bottom face to sit on the motor support bar as on the drawing and is slotted for belt adjustment. A steel plate, the bracket clamp plate, holds everything firmly, being much better than washers. The distance block, also of cast iron, is nominally shown as 1⅛" thick but this should be checked and made to a thickness that will bring the belt truly in line.

For the back centre support arm, the clamping bolts are made first and the holes for these, after carefully marking out, are drilled and reamed and the clamping bolts held in place temporarily for boring by a 4BA grub screw. The back of the casting is first faced in the four-jaw chuck, then clamped against the faceplate and the 1" hole bored to a nice sliding fit on the bar support. The ½" hole is not drilled until the whole machine is finished. It is then set up on the vertical slide, the back centre support arm in place on the bar support, a large centre drill is mounted in a collet in the machine, and while the back-centre support is left free on its bar, the support is prevented from swinging by two jaws of the four-jaw chuck. The saddle is then moved along and a deep centre made in the support casting, truly on the correct centre distance. It only remains to set the support arm back on the faceplate running true to dial gauge on this large centre hole, when the hole can be bored and reamed ½" dia.

The bar support bracket should require no explanation; it is set on the last tee slot of the cross-slide which brings the high side of the angle flush with the end of the cross-slide so that the arms can go down beyond the end. I finished the face of the short angle to surface plate. The photographs show how it is used. It forms a strong bridge which takes all angular strain off the vertical slide. The little box on the motor which can be seen in the photographs I made from sheet aluminium and contains a three-terminal connector and forms a connection box for the flexible cable and an easy means of reversing the motor. I have drawn two milling arbors for 1" dia. cutters as a suggestion and the machine will take a maximum of 4" dia. cutters with the bar support in place. A 3" dia. × ³⁄₁₆" cutter can be used to the full depth in b.m.s. with no trouble, cutting at a good speed. The speed is a little too low for end milling, the largest collet which can be taken is ⅝", which is not too bad on the fastest speeds but for the smaller end mills I find that I have sufficient power to take the larger pulley for the motor for the smaller end mills. I will therefore make a pulley of 2", 1¾" and 1½" and feel sure there will be ample torque at this speed; a longer belt will of course be necessary.

When milling gears, splined shafts, keyways, etc., it is necessary to lock the cross-slide. As no provision is made for this on the Super 7, I drilled and tapped two ³⁄₁₆" Whit. holes in line with the cross-slide gib screws and in the centre of each pair and fitted ³⁄₁₆" Allen cap screws. This saves altering carefully adjusted gib screws. The same was done with both vertical slides. Very little explanation should be necessary. The key set in the back of the main body is of course ⅜" square to fit the tee slots in the vertical slides. It is fitted in one piece and three No. 4 Allen cap screws are fitted flush with the surface of the key as shown in my drawing. A drill is then passed through the tee bolt holes and finally all burrs and sharp edges are removed.

When cutting gear teeth the indexing attachment is used and even for cutting a straight keyway I use it, as it holds the work completely free From movement and it takes but a few seconds to fit. For cutting bevel gears the vertical slide is mounted directly on the topslide of the lathe. This topslide is set to the cutting angle. The milling attachment is set with its spindle in a vertical position with the cutter centre right on the lathe centre. For this purpose the key in the back of the body is removed, but the two tee bolts will fit two adjacent tee slots and hold quite securely against cutting pressure.

After setting the cutter to the proper depth of cut, the saddle, cross-slide and vertical slide are all locked. After cutting down the centre of each tooth space the 'blank roll' is indexed (generally a $1/4$ pitch) with the indexing head and the 'offset' is made by the necessary amount of movement of the vertical slide. Bevel gears are thus simple to make, and are actually more rigidly supported than in a milling machine, where the indexing head has to be tilted upwards at difficult angles.

For such operations as gashing worm wheels, the universal slide is necessary as the spindle must be set at the correct helix angle. After gashing, the spindle is set back at right-angles for hobbing. For such things as cutting milling teeth in facing and angle cutters, the topslide is used as for bevel gears, but the base plates can be used for dog clutches which are cut with slotting cutters straight across. If these teeth are cut right across both sides with a narrower cutter than the groove desired and then indexed 180 deg., a perfectly radial slot is the result which will mesh without misalignment.

Adjusting the saddle

I had a little difficulty with my Super 7 when I first used it. No matter how carefully I adjusted the saddle gib screws, if the saddle was not too tight there was quite a lot of play in it. When the play was taken up, the saddle was almost immovable. Upon dismantling, I could see nothing wrong, but I noticed that the two outer gib screws which are supposed to hold the gib piece in place had rounded ends which fitted in rounded indentations in the gib piece. It was obvious that the gib was riding along a little and tightening up the adjustment as soon as the saddle was moved. When the saddle was stationary the gib piece shook into place and the saddle was loose again. I cured it by drilling and reaming a $1/8$" dowel pin hole right through the centre of the saddle and through the gib strip. A little peening of the inside face around the hole with a rivet punch prevented the dowel from working inwards against the bed and after fitting the dowel a fraction short, a little peening on the outside of the hole prevented it from working out.

Upon trying the lathe again I was amazed at the difference it made. I could now turn to the full capacity the belts could take, with no sign of chatter. I was so pleased that I treated the topslide in the same way and now I make all my machine slides with a dowel pin to hold the gib piece firmly in place.

CHAPTER 2

An Indexing Attachment for the Headstock

This indexing fixture has several advantages over the more usual fixture where a change wheel is mounted on the end of the headstock mandrel. No dismantling or changing over of any part of the lathe is necessary. It can easily be set up in ten seconds and removed in less than fifteen seconds. It is rigidly supported on three widely spaced points and is firmly held in all planes. It is instantly adjustable for wear and is highly accurate. If the Myford indexing head is owned, many of these parts can be used. Finally, it is not difficult to make.

Part E is supported on two pointed screws across the inside of the sloping belt guard and square with the lathe mandrel. Part A is set at the helix angle of the 16 D.P. worm and is held down into mesh with the lathe bull gear, which is 60 T. 16 D.P., by the swinging rod part F into the latch part G mounted on the front of the lathe headstock.

Commence with part A which is $9^9/_{16}$" of 1" $\times$ $^1/_2$" b.m.s. Drill for 2BA Allen screws $^3/_{16}$" dia., and counterbore with a flat-ended drill $^5/_{16}$" dia. for a depth of $^3/_{16}$", using an ordinary drill first of course. The oil holes can be drilled and also the $^3/_{16}$" dowel pin hole to secure part E at the correct angle.

Parts B and two off part C are next. These are made from cast iron rod from stock from any foundry. Part B is from $1^3/_4$" rod held offset in the four-jaw chuck $^1/_4$", to turn the 1" length to $1^3/_8$" dia. Do not bore but cut off at about $2^3/_{16}$" long. Part C, two off, are from $1^1/_2$" rod turned to $1^3/_8$" and faced off each end to 1" long.

Face off all three pieces in the four-jaw

The indexing attachment fitted to the lathe.

21

The indexing and milling attachments used in combination.

chuck along one side to 1$\frac{1}{8}$" from the finished diameter. Mill or shape a groove along this face to fit closely the bar A to a depth of $\frac{1}{8}$" bringing the thickness from the milled groove to the opposite side to 1". Bolt a short piece of the 1" × $\frac{1}{2}$" bar dead square across an angle-plate and mount the angle-plate on the faceplate in the lathe so that the face of the bar is $\frac{1}{4}$" off centre and its two edges are central using the cross-slide index to ensure this. Now clamp each part on this short piece of bar for boring. Parts C are merely centred and drilled $\frac{15}{32}$" and bored 2 or 3 thou under $\frac{1}{2}$" and then reamed $\frac{1}{2}$". Incidentally, I have noted that many articles in model engineering magazines show hand reamers being used in the lathe for reaming. These are quite unsuitable and the ordinary taper shank, spiral flute machine reamer should be used. Use a very slow speed in back gear and a fast feed straight in and straight out again. Two or three thou for reaming is plenty. Part B is set up in the same way: it is faced, centred and then turned to 1$\frac{3}{8}$" dia. to leave 1" long of the original and then turned accurately to 0.625" to leave $\frac{3}{8}$" shoulder, the 0.625" length being $\frac{3}{4}$". Drill $\frac{11}{32}$" and bore to a couple of thou under $\frac{3}{8}$" and ream $\frac{3}{8}$". Do not mount on part A until the worm shaft is completed.

A piece of $\frac{7}{8}$" b.m.s. just over 9$\frac{3}{8}$" long

is centred each end and faced using the fixed steady; for 3" or 4" from one end, take a couple of thou cut just to make sure the surface is running dead between centres. Set up the travelling steady over this part and make two grooves $\frac{5}{8}$" dia. and about $\frac{1}{4}$" wide at 6$\frac{9}{16}$" and 8$\frac{1}{4}$" from the other end to clear the tool for screw cutting. It is better to leave the bar the full diameter while screw cutting the worm.

If you do not have a tool for 29 deg. included Acme or worm, make one, using a 3" length of 1" × $\frac{3}{8}$" b.m.s. and at a height of $\frac{1}{2}$" from one end drill downwards along the length $\frac{1}{4}$" hole so that this hole breaks through the bottom about $\frac{1}{4}$" from the other end of the bar. This gives a good top rake to the tool. The tool is a length of $\frac{1}{4}$" round tool steel and the $\frac{1}{4}$" Allen screw to hold it is drilled from the top at right angles to the tool. Do not grind a flat on the tool as it can then be set for left and right-hand threads to suit any job on hand.

The tool is ground to 29 deg. included angle with a point $\frac{1}{32}$" wide; it is then perfect for 10 t.p.i. Acme also. I made another tool exactly the same but with a tool ground at 55 deg. included angle and now use this exclusively for screw cutting. It is always at the correct height and instantly available.

Set this tool right under the centre of the travelling steady jaws. If you do this you will

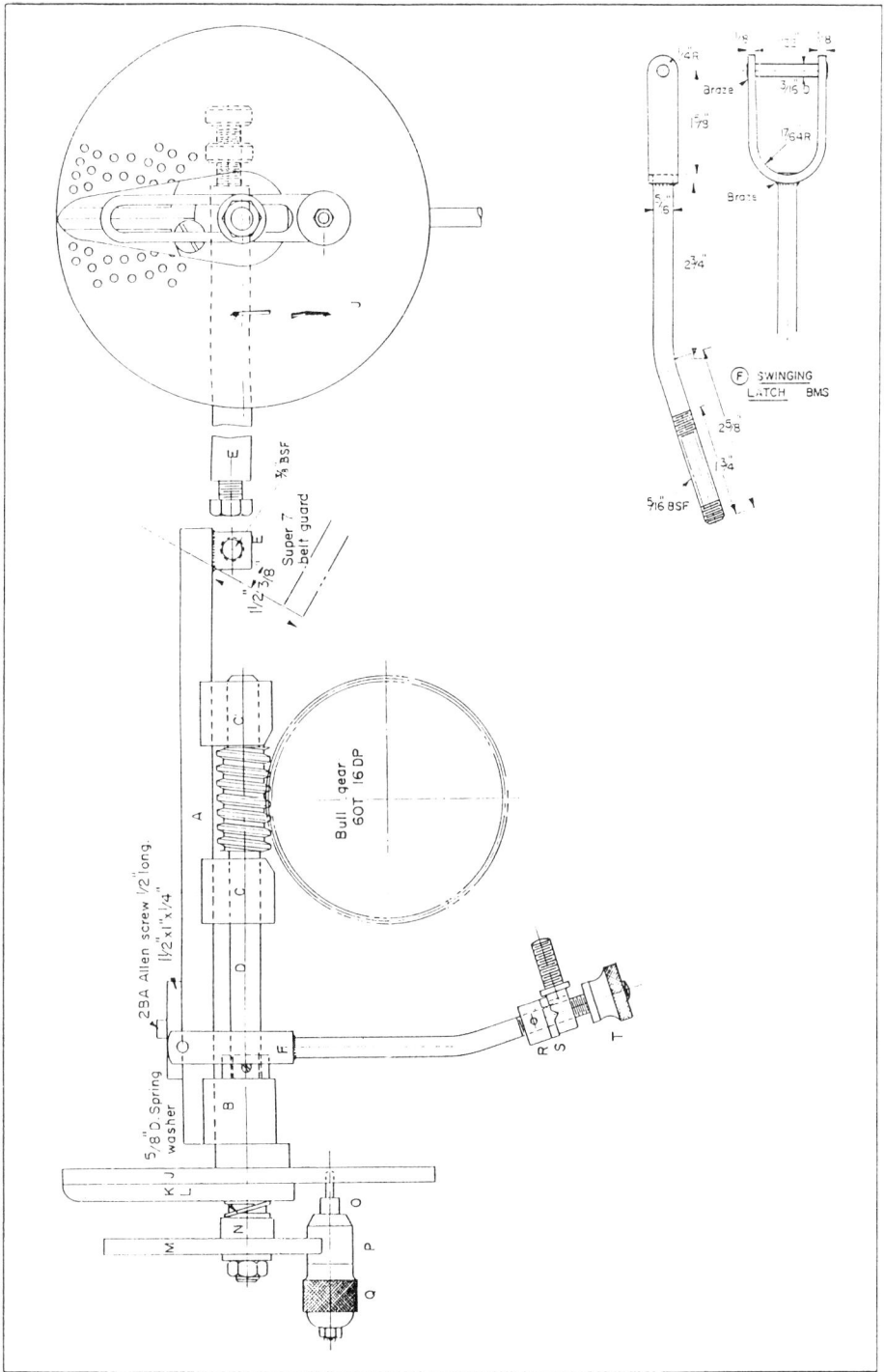

Bull gear
60T 16 DP

⅜ BSF

Super
belt guard

1½ 2 ⅜"

2BA Allen screw ½" long.
1½"x1"x¼"

⅝" D Spring
washer

¼R
Braze
1⅞
3/16 D
17/64R

Braze
⅝
6
2¾
Braze

F SWINGING
LATCH BMS

2⅝
3¾
5/16 BSF

A
B
C
D
E
F
J
K
L
M
N
O
P
Q
R
S
T

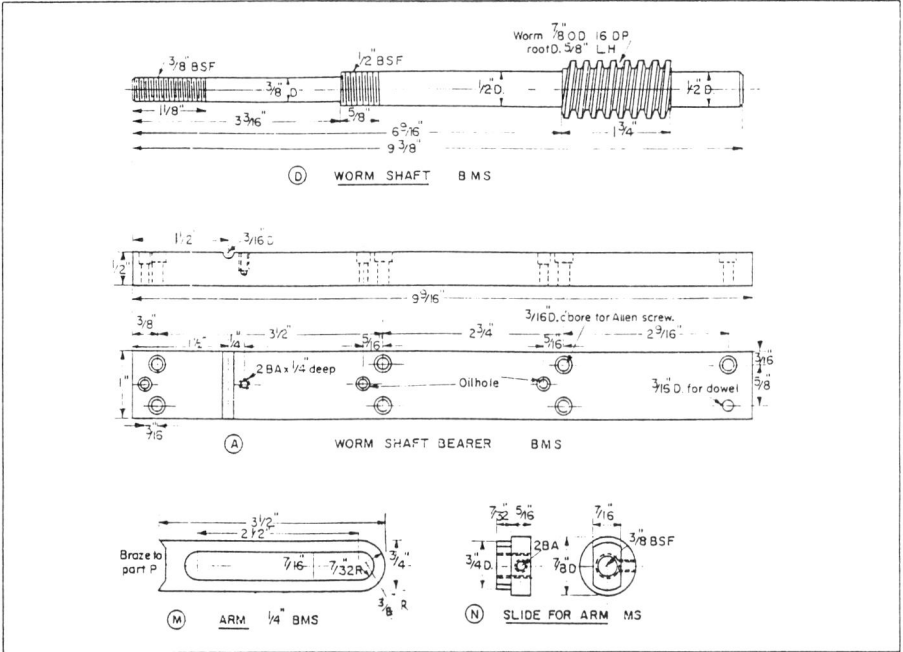

find that you can take enormous cuts with no sign of chatter and still obtain an excellent finish. The cutter is set for left-hand turning. Your handbook will tell you the gears to use for 16 D.P., which is the largest D.P. possible. Using back gear, and of course travelling from the headstock outwards, the worm is cut until the lands are the same width as the bottom of the grooves and root diameter is $5/8$". The topslide is used to get the correct width of groove. Next complete the turning of the worm shaft with the journals a nice fit free but without play in parts B and C. Screw cut for collar $1/2$" BSF and fit the collar face at $8^{1}/8$" from other end; fit a 4BA Allen grub screw and take a truing cut on the shoulder face. It is very important that this face runs true. The $3/8$" BSF thread on the end can now be screw-cut to $1^{1}/8$" long, to fit a standard $3/8$" nut.

Set up the bearings with the worm shaft

in place on part A and spot drill the bearings from part A. Do not drill deeper than $7/16$" No. 24 for tapping 2BA and make sure that you have 1" protruding from part B and that the worm has clearance each side of the two parts C. The side thrust of the shaft is taken between the collar on the shaft and part N and the worm should have side clearance.

When parts A, B, C, D are assembled, make part E, which is a $5^{3}/8$" length of $5/8$" square b.m.s. deeply centred truly at each end and faced. This is screwed with the one Allen screw to part A only. Lay the assembly on the bull-gear moving the worm shaft a few degrees to the left to mesh the worm firmly in the gear and at the same time bringing part E square with the headstock. Now at $3/8$" from the top face of the sloping belt guard, a position will be found for pointed screws H and I which will bring the worm central on the bullgear; at the same time note that the length of these screws will be correct for your particular lathe to bring the worm central longitudinally as well. In my case the position of the screws was $1\frac{1}{2}$" from the machined face of the guard. If all is well, make and fit the two pointed screws, one of which, the left one, has a hexagon head and is not touched once in place; the other is knurled and has a locking knurled nut. These are both $\frac{3}{8}$" BSF and the belt guard is drilled and tapped for these screws. The screws are case hardened on the points.

The swinging latch part F is made from a piece of $\frac{1}{2}$" × $\frac{1}{8}$" b.m.s. and a length of $\frac{5}{16}$" dia. rod. It is held in place with a piece of 1" × $\frac{1}{4}$" b.m.s. clamped on top of part A by one 2BA Allen screw and the $\frac{3}{16}$" hole for the $\frac{3}{16}$"

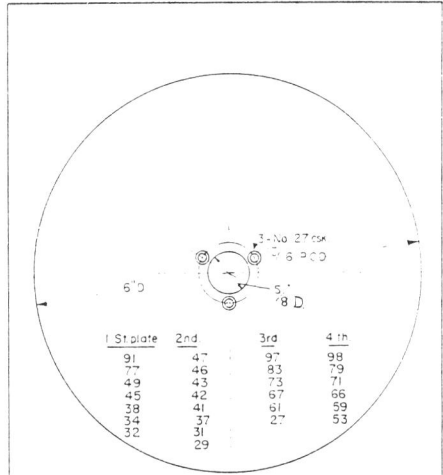

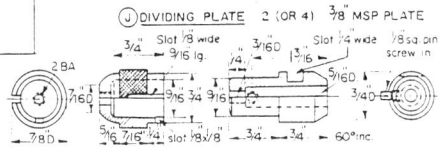

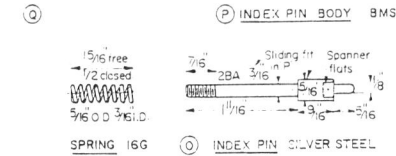

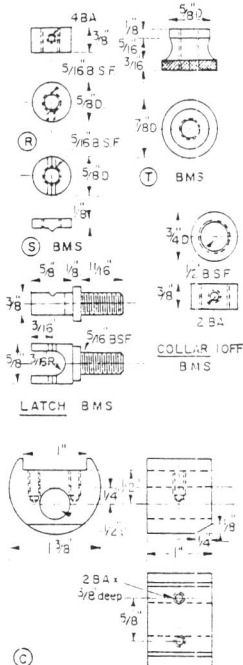

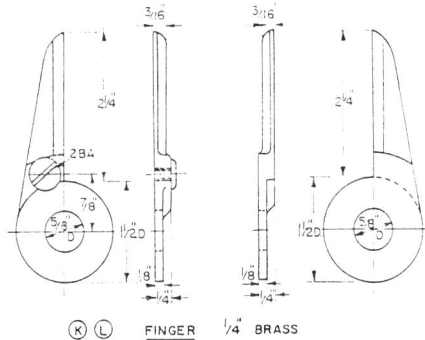

1st plate	2nd	3rd	4th
91	47	97	98
77	46	83	79
49	43	73	71
45	42	67	66
38	41	61	59
34	37	27	53
32	31		
	29		

(J) DIVIDING PLATE   2 (OR 4) $\frac{3}{8}$" MSP PLATE

(Q)   (P) INDEX PIN BODY   BMS

SPRING 16G   (O) INDEX PIN SILVER STEEL

(R)

(T) BMS

(S) BMS

LATCH BMS

COLLAR 1 OFF
BMS

(C) WORM SHAFT SUPPORT BLOCK
2 OFF   CI

(K) (L)   FINGER $\frac{1}{4}$" BRASS

25

pin is drilled half way into this piece and part A, interposing a piece of heavy paper while drilling so that end play in the pin can be taken up. I merely brazed the pin into part F and also the $^5/_{16}$" rod as being the easiest way. Also, after part F is finally secured in place and dowelled, this dowel should be short and not drilled into the centre hole. I also brazed along the bottom corner of part A to part E.

The latch part G is mounted on to the front of the headstock after bending the $^5/_{16}$" rod to come vertical in both planes. In my case it came vertically under the first line of the D in 'Myford' as on my sketch. It is tapped $^5/_{16}$" BSF and is also secured with a nut inside to lock it. The collar R is the vertical adjustment for the worm and bears against a $^1/_8$" thick washer with a Vee protrusion to engage in V-grooves in latch G. The holding nut T is knurled and holds the worm down in place to a proper mesh with the bull-gear. Part R is secured against movement once the adjustment is made with a 4BA grub screw.

Part N should now be made from b.m.s. $^7/_8$" dia. It is tapped truly, or better still screw-cut, and then tapped to fit the end of part D and two flats are milled or carefully filed to $^7/_{16}$" width on the $^3/_4$" dia. which is $^7/_{32}$" wide. A 2BA Allen grub screw is fitted with the end flattened and a small copper pad interposed between the screw and the thread on the shaft. This part N adjusts end play in the worm shaft. The larger face must be true with the thread.

Part P is the index pin body and should be turned from b.m.s. held in the three-jaw chuck and all turning, boring and reaming done from one setting before parting off. A left-hand turning tool should be used for turning the $^9/_{16}$" dia. Both the $^5/_{16}$" and $^3/_{16}$" holes should be bored and then reamed. The end is 60 deg. included and the $^1/_4$" groove is better filed before parting off; it must be square. The pin is $1^3/_{16}$" from the bevelled face and is tapped 2BA (blind hole) and the pin tightly screwed in and then filed to $^1/_8$" square.

Part Q is turned from b.m.s. and is bored a free fit for the $^9/_{16}$" end of part P to a depth of $^{13}/_{16}$".

The small hole in the end is tapped 2BA using the tailstock chuck while still in position in the chuck and before parting off. The outside diameter is knurled and the end domed. A $^1/_8$" groove $^9/_{16}$" deep is filed on one side only and at right angles to this groove another one only $^1/_8$" deep is filed. The land between these two grooves is reduced in height slightly so that on assembly it is possible to turn part Q only a quarter of a turn. This saves a lot of fiddling around when indexing. The index pin part O is made from silver steel and has a 2BA thread $^7/_{16}$" long on the end. The $^5/_{16}$" and $^3/_{16}$" parts should be a nice sliding fit in part P as accuracy of indexing depends a lot on this fit. A couple of small flats are filed on the end of the $^5/_{16}$" part to engage a small spanner and the $^1/_8$" part should be made to closely fit a $^1/_8$" drilled hole using the same drill you intend to use to make the index plates. The index spring should be quite strong and the length is important; it should allow the end of the pin to retract up to $^5/_{32}$" from the end of part P and no farther. The end of the pin is smoothly rounded and is hardened in oil and tempered to a medium straw.

The arm M is made from $^3/_4$" × $^1/_4$" b.m.s. and should be cut to length only after the slot is milled. Two $^7/_{16}$" holes are drilled at $2^1/_2$" centres and then a row of $^3/_8$" holes drilled along and close together. It is set up on the vertical slide and clamped at each end with suitable packing under it to clear the slide and then milled to $^7/_{16}$" wide using any suitable end mill up to $^7/_{16}$". Feed against the direction of rotation of course. Finally cut to length as shown on the drawing and round the end to $^3/_8$" radius. This part is then brazed into the body of part P, making sure you get it quite square.

The fingers parts K and L are made from $^1/_4$" sheet brass. Set up an ample area on the faceplate and face down a $1^1/_2$" dia. part on each to $^1/_8$" thick and bore carefully to $^5/_8$"

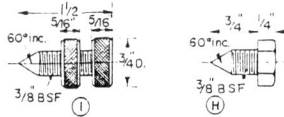

SUPPORT BAR FIXING PINS   BMS

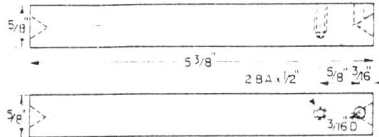

(E)   FIXTURE SUPPORT BAR   BMS

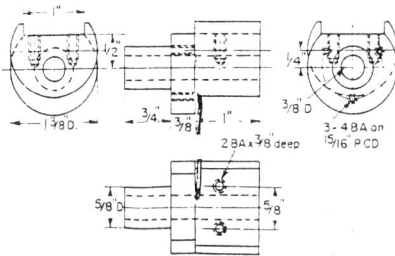

INDEX PLATE SUPPORT BLOCK   (OFF C)

Above and below: Some general views of the indexing and milling attachments in use.

dia. It is then cut with the hacksaw and filed to shape as drawing. K is flat on the back nearest the dividing plate and is fitted first. It has a large headed 2BA screw for clamping part L. This part is recessed on the back. The radial edges of both are bevelled to $^1/_{16}$" thick and are reduced on the arms only to $^3/_{16}$" thickness.

It only remains now to make the dividing plates and these are not at all difficult. Two plates will cut 85 per cent of every number

up to 100 but four plates are required to cut every number.

You will find I think that the two plates will fill all your needs, but in any case, ask your nearest engineering establishment to flame cut two (or four) discs from $^3/_8$" boiler plate $6^1/_4$" dia. Grip each one by $^1/_8$" of the edge in the chuck and face one side and turn as far as the chuck jaws to 6" dia. Set up the reverse side and get the edge of the plates running true and face the other side

Another view of the indexing and milling attachments in use.

until the plate is $1/4$" thick and of even thickness to the micrometer. Do not bore as yet. When all the plates are finished as above, make a drill bush from a piece of 1" × $3/8$" material, the idea being to set up this drill bush in the toolpost close to the dividing plate blank for drilling. The piece of steel is gripped under the toolpost and is centre drilled with the centre drill held in the chuck. Follow this with a $7/64$" drill and then a new $1/8$" drill carefully followed through to size the hole. This drill bush is then slightly centre drilled to chamfer on the side nearest the chuck and then heavily on the reverse side. Case harden the drill bush and then make a thick washer $1/4$" thick about $1 1/2$" dia. Scribe a $15/16$" diameter on this and carefully space off three equidistant positions and drill No. 24. Use this as a drill jig and drill and tap 4BA the three holes in the face of part 8. The hole in this washer is $5/8$".

Now procure a piece of $1/4$" plywood about two or three feet square; make sure it is flat. Set up the indexing fixture on the lathe and mount the plywood with its retaining washer just described to clamp it firmly against part 8. Assemble all parts except index fingers and index pin, but including parts N and P. Clamp a piece of steel strip about 1" × $1/4$" or $5/16$" on to the arm of part P and long enough to scribe a circle around near the edge of the plywood. Clamp a scriber or pointer to the end of this strip and scribe a circle as large as you can.

Calculate the circumference of this circle and divide this by 91, which will be your first ring of holes. Use dividers and go round this circle until you finish up at your starting place. You will probably have to go round several times until you get the dividers right.

Remove the whole indexing attachment – it takes only a few seconds – and set up the first plate in the chuck to run true radially and on the face. Centre and bore to exactly 0.625" using a plug gauge exactly the same size as the end of part B.

Now set up the hardened drilling jig under the toolpost, clamp dead square in every way, check by putting the $1/8$" drill in the tailstock drill chuck and set until the jig is free on the drill. Move the cross-slide outwards until the hole is $1/4$" from the edge of the plate, set close up against the plate but just clearing, and clamp both cross-slide and saddle.

An electric drill is now used to drill the holes in the plate using the new $1/8$" drill. Set up the indexing head with its temporary plywood index plate and turn the pointer to the first position. Do not go backwards if you pass the point except by going a long way backwards and approaching again. In this position mark this point on the plywood with a pencil and drill your first hole. Count 60 points round the circle and mark again and drill the second hole in this position. Drill each hole in the same way, counting 60 points for each indexing.

This may seem a crude way of indexing, but let us look at it. At 18" radius circle an error of 0.020" (which is a lot and it is quite easy to get much closer than this) would give an actual error of 0.003" if there were no other reduction. As it is, you have a reduction of this of $^1/_{60}$ by the worm reduction, which brings the original error of 0.020" to 0.00005". In addition, when you use this plate for milling a gear of 6" dia. you get down in theory to an error of 0.000008" or eight millionths of an inch! Your drilling will not approach this, so the accuracy is high.

After completion of this ring of holes, mark another circle a little closer to the centre of the plywood and set the drill bush $^1/_4$" closer to the centre by means of the cross-slide index; reclamp the cross-slide. The second ring is done in the same way but the circle this time is of 77 spaces and indexing is every 60th hole as before.

The circles on the first plate are 91, 77, 49, 45, 38, 34, 32, seven circles and on the second plate are 47, 46, 43, 42, 41, 37, 31, 29, eight circles. I would recommend that you also drill on the first plate an inside row of 25 holes as this is handy for indexing 125 divisions for an eight-thread leadscrew, indexing being 12/35.

The third plate if made is 97, 83, 73, 67, 61, 27, six circles and the fourth plate is 89, 79, 71, 66, 59, 53, six circles. In each case 60 spaces are counted off for each drilling.

In the chart, whole numbers are number of complete turns. Where no fraction is included, any one hole in any circle may be used. The denominator in the fractions is the index circle. The numerator is the number of holes spaced off on the circle. Spacing for numbers of one half or less of the circle is done inside the fingers and fingers are moved around for each indexing in the same direction as the index pin. For numbers above one half of the circle, the indexing is done over the top and round the back of the fingers and the fingers are moved in the opposite direction to the index pin.

Indexing chart for 60-1 worm reduction

No. Divisions	No. turns and circle	No. Divisions	No. turns and circle	Plate No.
1	60	51	1-6/34	
2	30	52	1-14/91	
3	20	53	1-7/53	4
4	15	54	1-5/54	
5	12			
6	10	56	1-3/42	
7	8-28/49	57	1-2/38	
8	7-16/32	58	1-1/29	
9	6-30/45	59	1-1/59	4
10	6	60	1	
11	5-35/77	61	60/61	3
12	5	62	30/31	
13	4-56/91	63	40/42	
14	4-14/49	64	30/32	
15	4	65	84/91	
16	3-24/32	66	70/77	
17	3-18/34	67	60/67	3
18	3-15/45	68	30/34	
19	3-6/38	69	40/46	
20	3	70	42/49	
21	2-42/49	71	60/71	4
22	2-42/49	72	35/42	
23	2-28/46	73	60/73	3
24	2-16/32	74	30/37	
25	2-18/45	75	36/45	
26	2-28/91	76	30/38	
27	2-10/45	77	60/77	
28	2-6/42	78	70/91	
29	2-2/29	79	60/79	4
30	2	80	24/32	
31	1-29/31	81	20/27	3
32	1-28/32	82	30/41	
33	1-63/77	83	60/83	3
34	1-26/34	84	35/49	
35	1-35/49	85	24/34	
36	1-30/45	86	30/43	
37	1-23/37	87	20/29	
38	1-22/38	88	45/66	4
39	1-49/91	89	60/89	4
40	1-16/32	90	30/45	
41	1-19/41	91	60/91	
42	1-21/49	92	30/46	
43	1-17/43	93	20/31	
44	1-28/77	94	30/47	
45	1-15/45	95	24/38	
46	1-44/46	96	20/32	
47	1-13/47	97	60/97	3
48	1-8/32	98	30/49	
49	1-11/49	99	40/66	4
50	1-9/45	100	27/45	

For angular dividing, 1 degree = 7/42.

CHAPTER 3

Ball-bearing Cone Centres

The two ball bearing cone centres described here will be found very useful for turning work which has been bored or for tubing, or for setting up hollow work to run true on the faceplate or in the four-jaw chuck. They will take any internal bore where the ordinary centre leaves off, up to 4" dia., and save the necessity to plug up the end or make end covers to take an ordinary centre.

Start by making the two patterns. These are simple turning jobs and the core print can be turned in one piece with the pattern or could be glued on true afterwards. The 3" dia. shoulder on the larger one and the $1\frac{1}{2}$" dia. shoulder on the smaller one are for holding the casting for boring. The castings should be made in grey iron.

It is possible to obtain two stock pieces from most foundries which will save the trouble of making patterns, but it is necessary to obtain solid pieces with no cored hole in this case and I think it is easier by far to make the patterns. The core print in the larger pattern is 1" dia. and $1\frac{5}{8}$" long and the core should be specified as being $3\frac{1}{4}$" long. The smaller pattern core print is $\frac{5}{8}$" dia. and the core should be specified as being 3" long, the print itself being $1\frac{1}{2}$" long.

Set up the castings in the four-jaw chuck by the largest diameters and turn true the 3" dia. and $1\frac{1}{2}$" dia. shoulders so that these will grip truly in the chuck. Each casting is now held truly in the four-jaw chuck on these turned shoulders and the boring is carried out after first facing off the large diameter face. In the case of the larger casting, bore first to 1.3779" to a depth of 1~" to a light tap fit on the 6202 rigid ball race; the tool can have a small radius as the ball race does not bottom on this hole, all side thrust being taken by the outer 6203Z ball race. Next bore to a depth of $1\frac{3}{16}$" to 1.380" dia. to clear the 6202 ball race. Finally bore to a depth of $\frac{9}{16}$" to 1.5748" to a LIGHT tap fit on the 6203Z ball race. The small groove for the circlip, whether 40mm or $1\frac{9}{16}$" internal type, should be made to clear the width of the ball race which is 0.4724". The 6203Z ball race is the type with one shield which should of course be fitted with the shield outside to hold in the grease and to keep out swarf.

Finally a centre is drilled in the bottom face of the hole and this is drilled $\frac{13}{64}$" and tapped $\frac{1}{4}$" Whit. for a $\frac{1}{4}$" Allen cap screw, the purpose of which will be explained later. The outer 4" dia. is now turned, not to a good finish necessarily as this and the cone face is later turned in position on its own bearings. The lightening recess is next turned, approx. $\frac{3}{4}$" wide by about $\frac{1}{2}$" deep, which improves the appearance.

The smaller casting is now set up on the turned shoulder in the same manner as the larger one, and the cored hole bored out to fit the R-5 ball race which is 0.7480" dia., to a depth of $1\frac{9}{16}$" with small radius as before. Bore to $1\frac{1}{8}$" dia., to a depth of $1\frac{3}{16}$", but leave the end face to an angle of about 30 deg. as drawing. Then bore to fit the 6202Z ball race, leaving a square shoulder to a depth of $\frac{9}{16}$", the diameter being 1.3779", making it a very light tap fit as before. The little groove for the circlip is now bored to either a 35mm or $1\frac{3}{8}$" internal circlip, leaving a space to fit the width of the ball race which is 0.4331". The shielded side of this race goes to the outside as before. The end of the hole is now centred and drilled and tapped $3/16$" Whit. for an Allen cap screw. The outer diameter is turned to just over $2^1/2$" and again a small recess in the back face can be turned as in the drawing for appearance sake.

Now set this trued up edge of each casting in the four-jaw chuck so that the turned shoulder runs true and also the machined face runs true against the chuck jaws; the machined shoulder can be turned off with the topslide set to 30 deg. and the turning taken to as close to the chuck jaws

as possible. Do not bother about a good finish at this stage as this 30 deg. face and also the large diameter edge is finish machined in place on its own bearings as will be explained later. At this setting face off the outer or smaller face to a good finish to a width of $1^7/8$" for both of the castings.

The two spindles are both made from $7/8$" b.m.s. Face off the larger one to $4^{13}/32$" long and the smaller one to $4^9/32$" and centre each end. Shoulder down the longer to square face $1^{17}/32$" long to diameter of 0.6693" a medium press fit on the 6203Z ball race and at 0.4724" groove for the 17mm (or $21/32$") external type circlip. Turn up

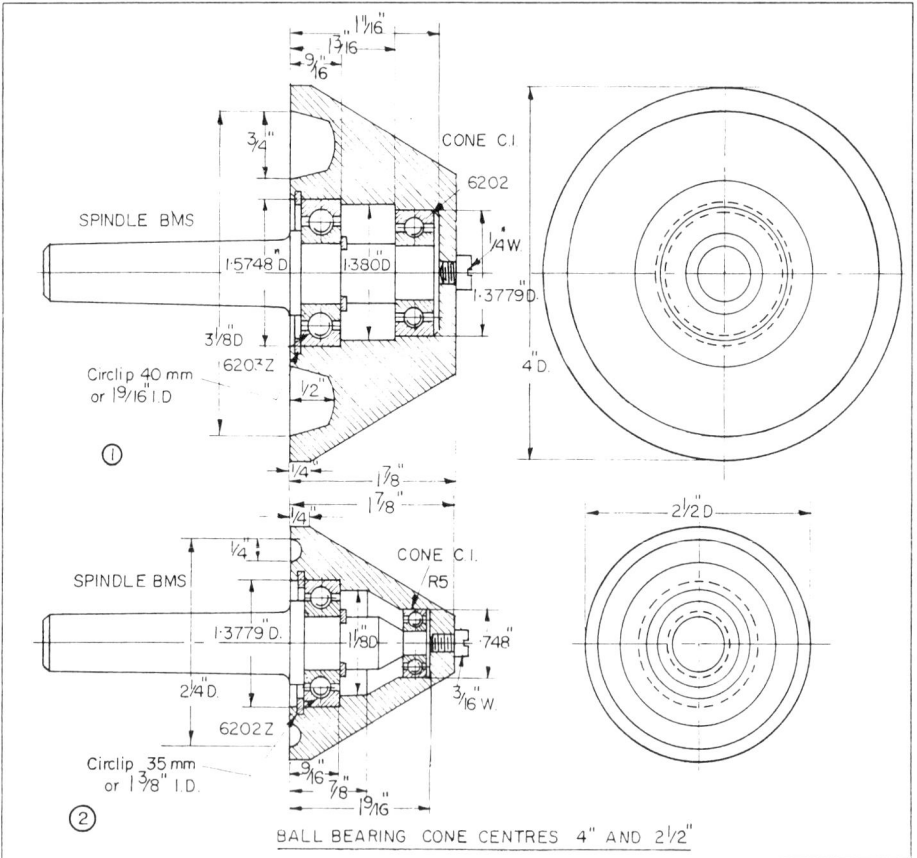

BALL BEARING CONE CENTRES 4" AND $2^1/2$"

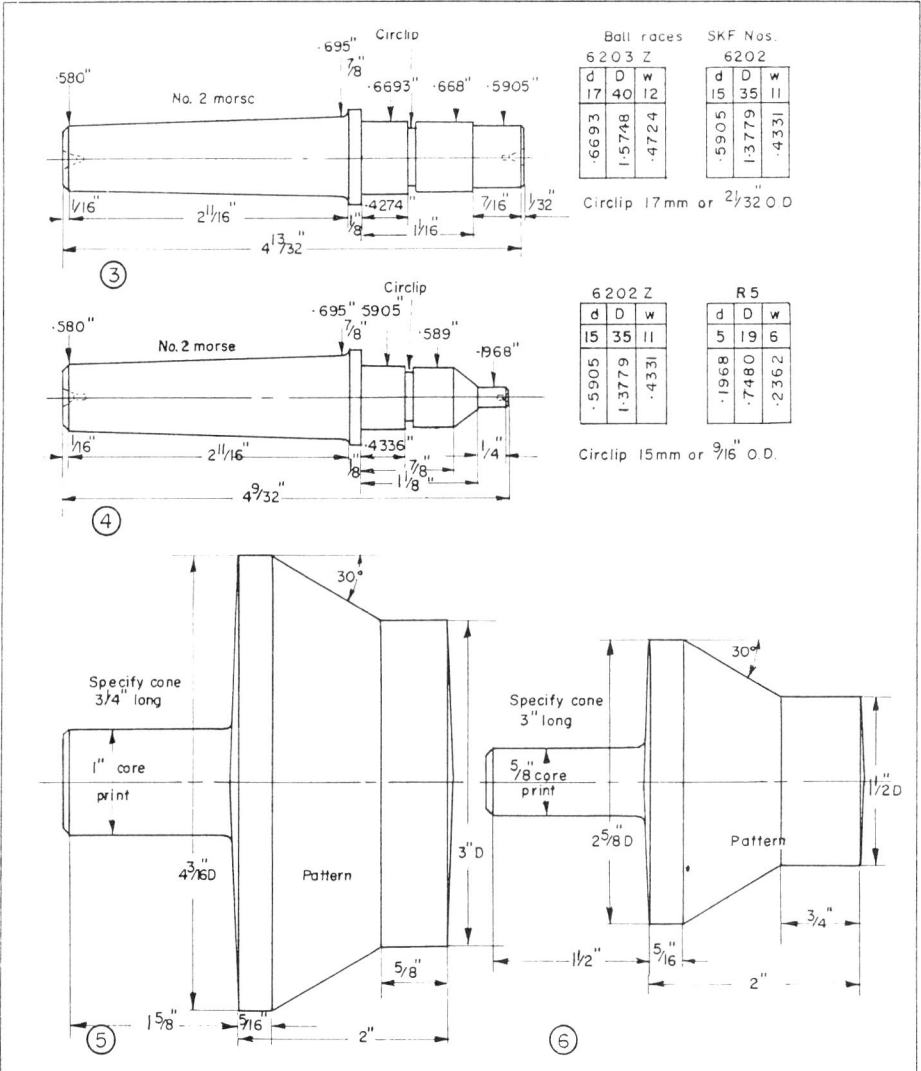

Ball races			SKF Nos.		
6203 Z			6202		
d	D	w	d	D	w
17	40	12	15	35	11
.6693	1.5748	.4724	.5905	1.3779	.4331

Circlip 17 mm or $\frac{21}{32}$" O D

6202 Z			R 5		
d	D	w	d	D	w
15	35	11	5	19	6
.5905	1.3779	.4331	.1968	.7480	.2362

Circlip 15 mm or $\frac{9}{16}$" O.D.

to this groove 0.668" which will clear the 6203Z ball race and then turn to 0.5905" to a shoulder at $1\frac{11}{16}$" from the $\frac{7}{8}$" shoulder for the 6202 ball race; the $\frac{1}{32}$" surplus on the end is for bevelling the edge.

The shorter spindle is machined in the same way to 0.5905" for a length of $1\frac{13}{32}$" and groove for circlip cut at 0.4331" from the shoulder to fit the 6203Z ball race; then turn to 0.589" dia. up to this groove, then to a diameter of 0.1968" up to $1\frac{1}{8}$" from the $\frac{7}{8}$" shoulder for the R-5 ball race. Turn the

33

shoulder down at 30 deg. to leave a small shoulder on the 0.1968" dia. Bevel the end. All that remains now is to turn the other ends of the spindles to No. 2 Morse taper leaving the $^1/_8$" wide shoulder at full $^7/_8$" dia. Grip the shafts in a carrier with soft pad on the 0.668" dia. and the 0.589" dia. in each case to revent burring on the ball race journals.

I have given the diameters at the small and large ends which apply in my case, but check these dimensions with a truly fitting centre of your own, so that the shoulder protrudes from your tailstock barrel about $^1/_4$". Three or four pencil lines along the length of the tapers and twisting the tapers no more than a quarter turn will check your angle setting in the tailstock barrel. Mark your tailstock centre on the top before removing for checking so that it is returned in the same position each time, as it could be out a little on the point, which would upset your angle setting.

When the spindles are completed to a good finish and accurate fitting, assemble the ball races and pack with light grease; do not forget the circlip. The assembly should now tap gently into place in the cones and the internal type circlip is then inserted to hold all securely.

Now cut off a $^1/_4$" Whit. Allen cap screw to a length to clear the end of the spindle for the large cone and the same but $^3/_{16}$" less for the smaller one and fit these two screws securely. Cut off a short piece of an Allen key to fit each screw say about $1^1/_2$" long

and grip the piece in the three-jaw chuck to protrude sufficiently to engage the Allen screw the full depth of the hexagon hole in screw head. Knock the larger of the cone centres into place in the tailstock barrel firmly and bring up to and engage the Allen screw over the piece of key in the chuck. The topslide can now be set over to 30 deg. which will clear the tailstock easily and the cone can be finished turned on the angle and on the outer shoulder to a good finish and with perfect accuracy. The same procedure is followed for the smaller cone centre. If during use the face becomes damaged at any time, it is a simple matter to return the angle to true it up and the $^1/_4$" wide shoulder width on the outer diameters will provide ample material for many true-up skims.

Drawings Nos. 1 and 2 are general assembly drawings with the cast-iron cone shown in section. Drawings Nos. 3 and 4 show the dimensions of the spindles only with a chart of the ball races dimensions. Drawings 5 and 6 show the pattern dimensions which show a slight taper on the faces of the cones for drawing from the sand. The cones are cast horizontally so that there will be no tendency for the cores to float upwards or move sideways. Consequently no taper is required on the core prints, which should be parallel and accurate. These two cone centres will be found quick and simple to make and are quite inexpensive.

CHAPTER 4

Gear Cutting in the Lathe

Most model engineers from time to time need to cut gear teeth; this means that an involute cutter of the correct pitch and number must be purchased for a one-off job, or else an attempt is made to make a fly cutter or multiple tooth cutter of approximately the correct shape from some existing gear. I have done this myself more than once, but the difficulty of backing off and getting the correct shape has left a lot to be desired in the finished gear, in fact in spite of every care taken the results have been so poor that the job has left me very unhappy.

The only solution seemed to make a complete gear hobbing machine using my Myford Super 7 as a base, and to make my own gear hobs. A great deal of thought went into means of backing off a spiral gear hob where the pitch of the hob lies in a line along the helix angle and not along the centre line. It is possible to cut this pitch using the taper turning attachment and setting over the tailstock to turn parallel, but when it comes to backing off correctly along this curve, by milling or turning, the difficulties mount up enormously, and unless adequate backing off is done accurately, the hob would be useless.

After much thought, I decided to make a 20 D.P. hob not with spiral teeth, but with circular lands and backed off accurately with a milling cutter, backed off too in a straight line rather than in a circular fashion. I designed a fixture to hold the hob for backing off and after hardening the h.s.s. hob and grinding on the front face of the teeth, I tried it out by making a 46 T and

25 T change wheel which can be used on the Myford Lathe.

The results were first class. One hob 20 D.P. will, of course, cut gears accurately of any number of teeth from the smallest to a rack, more accurately in fact than a full set of eight cutters in that pitch. On examination of the teeth, the curves, thickness and finish appeared perfect. Under a low powered glass it was possible to see that the curve was in fact a series of very small flats, and the gears when run together were in fact a little noisier than gears hobbed in the true manner, but after an hour or so of running in, no difference was apparent. In cutting these teeth I should have stated that I indexed around a second time by starting off the second time with a gear indexed a half tooth and the cutter set along horizontally one half of the pitch. This was necessary, particularly in the smaller gear, in order to do the necessary undercutting of the base of the teeth.

I now intend to make a hob of 16 D.P., 24 D.P., 32 D.P. and 40 D.P. this with the 20 D.P. cutter will enable me to cut gears of any number of teeth in all of these pitches. These five hobs will do the same work of forty involute cutters, which would cost to purchase more than the lathe. The five hobs can be made for a few shillings in high speed steel and, while a great deal of accuracy is required, it is straight line machining which can all be done on the lathe or in the milling machine if the reader is fortunate enough to possess one.

The first thing to make is a small pattern for the backing off fixture and two small patterns for the caps. Leave a good $^1/_{16}$" for machining on the top and bottom and the two ends of the main casting and the same amount on the top and bottom and one side of the two caps. I left the cap for the small end flat on its underside but left a half hole in the underside of the large hole end, about $^3/_4$" dia. with a corresponding half hole in the main pattern. The width across the top of the central gap of the main casting is $2^1/_8$" and the bottom width is $1^5/_8$" to give plenty of taper for draw. The ends are only slightly tapered inside with practically no taper on the outside at all.

Face off the main casting on top and bottom to $1^1/_8$" thickness and the top and bottom faces of the two caps to 1" thickness. Scrape these faces to the surface plate. Four $1^1/_2$" × $^5/_{16}$" BSF cap screws are required for the two caps and these are

clamped and drilled and tapped in place and the heads are recessed flush into the two caps. The outer faces of caps and main casting are brought flush.

Drill the four $^3/_8$" holes for the tee bolts at $3^1/_8$" centres both ways, make sure these four holes are square with the length of the casting and with each other. Mill a $^3/_8$" keyway $^1/_4$" deep along the centre of the rear two $^3/_8$" holes as seen in the end view drawing. Fit the $^3/_8$" square key in the groove and secure in place with three 4BA cap screws as seen in side elevation. The $^3/_8$" drill can now be passed through the casting and through the key which will cut it into three pieces. Remove all burrs and replace the key permanently. Set up the casting on the cross-slide and face off each end to leave it $4^1/_4$" long. Round off the corners for appearance.

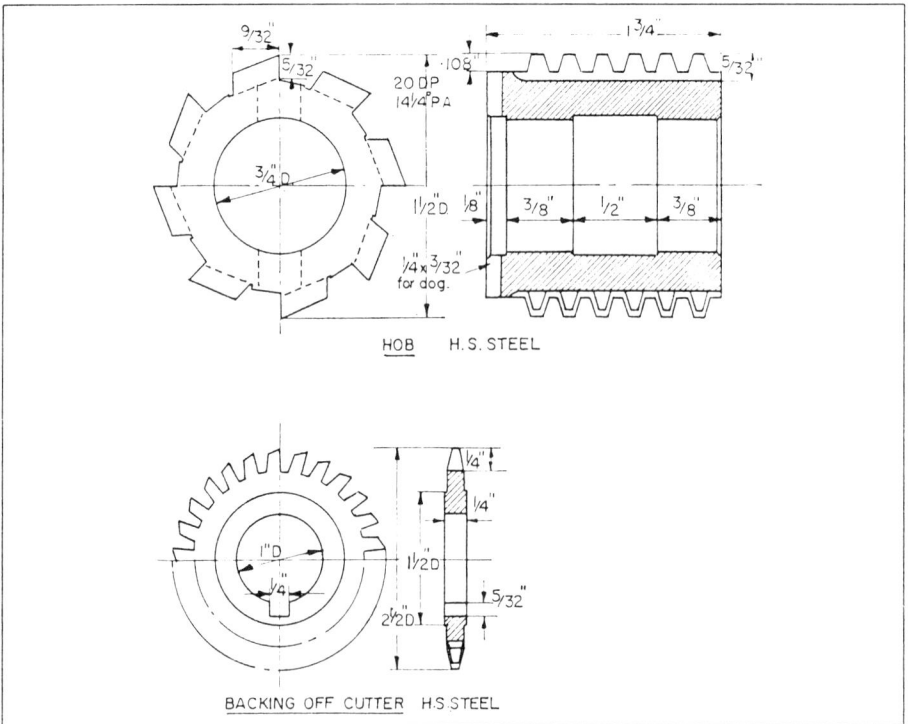

HOB   H.S. STEEL

BACKING OFF CUTTER   H.S. STEEL

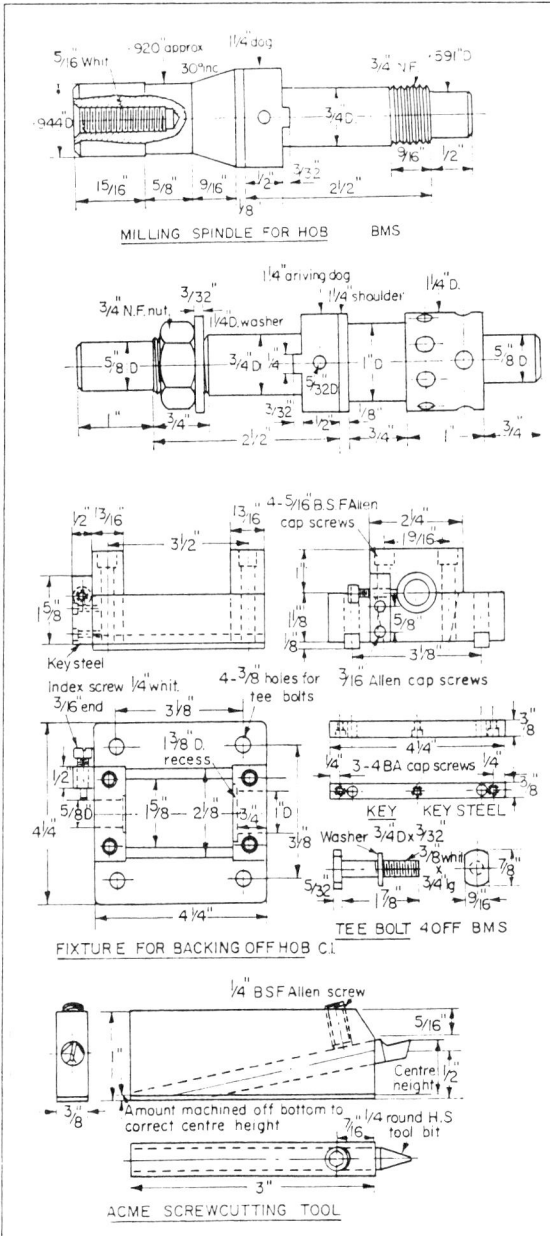

MILLING SPINDLE FOR HOB   BMS

FIXTURE FOR BACKING OFF HOB C.I

ACME SCREWCUTTING TOOL

For boring, I set up my elevating heads and my quartering table as seen in one of the photographs, but this can be done by setting up on the vertical slide or packing up on the cross-slide. However, first make the four tee slot bolts and washers from $^7/_8$" b.m.s. and the washers from $^3/_4$" steel. Also see that you have an accurate $^5/_8$" dia. boring bar of suitable length. Insert a piece of 0.010" drawing paper between the two caps and the main casting before boring and drill the small end $^9/_{16}$"; set a small boring tool in the chuck and bore out accurately to a gauge $^5/_8$"; this should be right on the joint line and in the centre of the casting.

The $^5/_8$'" boring bar can now be run in this hole which will stiffen it up for boring the 1" hole in the other end. This should also be bored to gauge and is recessed inside to $1^3/_8$" dia. to leave the journal $^3/_4$" thick.

The indexing block, made from $^1/_2$" key steel, is now made and fitted with two $^3/_{16}$" Allen cap screws with heads set in flush. The inside face of this must just clear the $1^1/_4$" dia. shoulder on the shaft and the indexing screw right on the joint line of the castings. The indexing screw is $^1/_4$" Whit. with $^7/_{16}$" dia. head and $^1/_4$" thick with slot for screwdriver. The point is $^3/_{16}$" dia. and $^1/_4$" long with rounded end and is case hardened on the end.

The next job is the indexing spindle to fit the journals and should be made to your gauges. A most necessary accuracy is the width of the 1" journal, as there must be absolutely no end movement to the spindle. This can be assured by repeatedly trying

37

Gashing the hob using a 2¹/₂" × ³/₁₆" slotting cutter.

the journal cap and measurement. The ³/₄" part is for the hob and must be made to size accurately, preferably to a gauge of, say, a piece of ³/₄" silver steel or stainless steel. The thread is ³/₄ NF which is 16 t.p.i. or some other fine thread could be used. It must be screw-cut. The driving dog is made from a piece of 1¹/₄" b.m.s. and should be milled right across while held in the chuck and then indexed 180 deg. and milled right across again. The tongue is then measured and the necessary adjustment made to the vertical slide and milled right across again and at 180 deg. This ensures that the tongue is dead central. It is ¹/₄" wide × ³/₃₂" deep and the body left is ¹/₂" thick. *Do nor pin this dog until after the hob is made and set up.*

Set up the spindle temporarily in the bottom casting and scribe a line around the 1¹/₄" dia. part through the centre of the indexing pin hole using a drill or other means, this is to ensure that the eight indexing holes line up with the indexing screw. Set up the spindle truly in the collet or other chuck with the 1¹/₄" part next the four tommy-bar holes in this 1¹/₄" dia.

journal. The tommy-bar holes are set at halfway or ¹/₁₆ indexing to the ³/₁₆ indexing holes and are approximately at ¹/₄" centre from the other edge.

The ⁵/₈" dia. end next to this 1¹/₄" dia. journal is for holding in the chuck (I used a ⁵/₈" collet chuck) while gashing the hob; this saves making a special mandrel for this part of the job. The actual drilling of the indexing and tommy-bar holes is done with the electric drill through a piece of ³/₄" key steel set in under the toolpost with a hole dead on the centre-line, drilled in place with a centre drill and ³/₁₆" drill held in the chuck of the lathe. If the point of the drill is thinned, no trouble will be experienced drilling with the electric drill.

I made the rack cutter next, so that no delay would occur once the hob was ready to be set up for backing off. This rack cutter is used for backing off the hob and must be accurately made to 29 deg. included angle. If hobs of 20 deg. pressure angle should be preferred, they can be done just as easily but in this case an angle of 40 deg. included would be required for the backing off cutter. Each side must be 14¹/₂ deg. or the gear

teeth will not be square or in line with the centre. I used a piece of $2^1/_2$" dia. h.s.s. and bored to 1" dia. with a $^1/_4$" × $^1/_8$" keyway. The cutter was made $^1/_4$" wide reduced to $^3/_{16}$" leaving a shoulder each side at $1^1/_2$" dia. the full width. After turning to $14^1/_2$ deg. each side, I gashed it for 24 teeth using a home-made cutter.

The depth of the gashing is $^1/_4$". A slight front rake to the teeth is about $^1/_{32}$". Side clearance on this cutter was first obtained by milling on the horizontal centre-line, with the milling head set at an angle to the vertical and using a small radial cutter.

The cutter was hardened by heating to red heat slowly and then quickly to yellow blistering heat with the oxy-acetylene torch and quickly plunging into oil: no tempering. After cleaning up, it was set up on an end mandrel again and ground on each side of the teeth using the toolpost grinder and the side of a small white grinding wheel. Finally the ends of the teeth were ground to the backing off angle and to correct the end width, which is most important.

To get this end width requires a gear tooth measuring instrument. I used a vernier caliper with a piece of bright mild steel dressed down to the tooth thickness on the pitch circle of a 20 D.P. gear which is 0.0785". A bit of 14 s.w.g. steel given a slight polish was just right. To determine the pitch line, it was necessary to make the bit of steel just 0.050" (the addendum of a 20 D.P. gear) short of the length of my vernier calipers. I did this by making sure I had the steel exactly the full length of the caliper jaws first and then measuring with the micrometer, filing off exactly 0.050" (dead square with one side, of course) and I had an accurate gauge. The ends of the backing off cutter teeth were then ground until this gauge just touched the sides of the $14^1/_2$ deg. teeth. This gave me the accurate width of the teeth, in relation to the correct depth. I should have made the smallest size of hob first and then progressed up to the larger ones, which would have meant grinding off only the ends of the teeth. As it is now, by making the 20 D.P. hob first, I will have to thin down the sides of this cutter to make the smaller hobs. I should have stated that the bit of gauge steel was clamped between the vernier caliper jaws, backing against the slide of the caliper, leaving an accurate gap for width and depth in the tips of the jaws.

I was now ready to make an actual start on the hob itself. A short length of $1^1/_2$" dia. annealed high speed steel was purchased. A piece $1^7/_{16}$" long was cut off in the saw, gripped in the 3-jaw chuck, faced, centred, drilled and bored to my $^3/_4$" gauge, a light push fit. The centre was relieved a few thou.

and also the outer end as per drawing. The shoulder was turned down to 1¼" for ⅛" length and the ¼" wide × ⅛" deep driving groove was milled in place using a ³/₁₆" slotting saw and indexing 180 deg. after milling right across, measuring and going over it again.

A short stub mandrel was then turned and the opposite end faced off truly to 1⅜" long. The hob was next mounted on the mandrel from the fixture between centres, a soft pad being used to protect the short ⅝" end. The outside dia. was then skimmed true; ¹/₁₆" was turned from each end to 1¼" dia. so that the teeth would be well inside the ends. The driving dog (unpinned) and the nut and washer is of course used for turning. I decided to first turn the teeth using my Acme tool in order to get an accurate depth for the teeth and then to back off only with the milling cutter. This simple tool is so useful that I have made a separate drawing of it and I also made two of them, one with the tool ground to 55 deg. for screw-cutting, which I use exclusively.

I found by measurement that I could get 7 teeth on the blank for 20 D.P. so was able to determine the position of my first groove. I ground the tool a good deal thinner than the cutter, so that I could feed it along with the topslide after going in some distance to clear the sides of the tool. The spacing of the grooves was done by a saddle stop and a thickness gauge of 0.1571" which is the circular pitch of 20 D.P. and also keeping a check on the saddle handwheel graduations, the readings for which I had worked out beforehand. By keeping the topslide handwheel graduation to the zero mark to the left each time and moving it only a predetermined amount, I was able to keep the depth and width of the grooves very accurate. The depth for 20 D.P. is 0.108" and the width was determined by using the backing-off cutter as a gauge.

Having formed these 7 teeth rings, the hob was gashed with eight grooves using a 2½" × ³/₁₆" slotting cutter, it was set about ¹/₃₂" to the rear of the vertical centre-line to give the hob some front rake, which improves cutting enormously; I milled to a depth of slightly more than ⁵/₃₂". At the same index

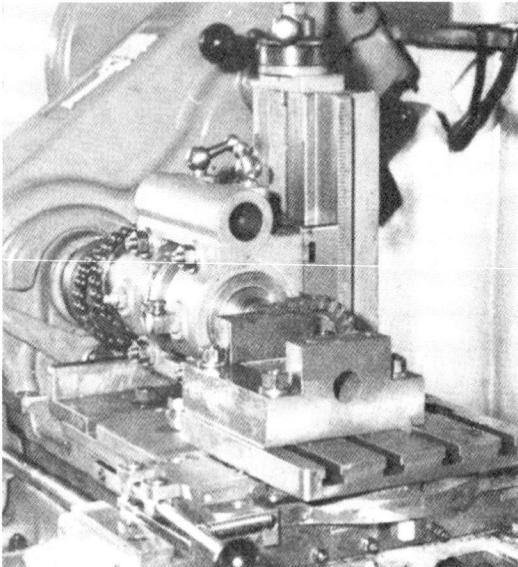

Facing the ends of the fixture casting on the quartering table using the elevating headstock.

The 'backing off' operation.

setting, the backs of the teeth were milled to just meet the edge of the slot. This left a tooth width of approximately $9/32$".

It remained now to mill the backing-off angle on these teeth. The backing-off fixture was set up on my quartering table and the milling spindle set up on the vertical adjustable heads, as seen in one of the photographs. The same thing could be done on the milling machine and also could be done by mounting on the vertical slide at the back of the cross-slide facing the operator.

It is necessary now to set up the hob so that the heel of the teeth is set to the highest point. This is done by slacking off the nut on the mandrel and with the index screw in place in one of the holes, turning the blank (and the driving dog) to the necessary position. In this position the cutting edge of the teeth is away from the operator and the heel of the teeth on the vertical centre-line. Tighten the nut in this position without allowing the hob to move. The pin through the driving dog and mandrel can be drilled for the pin, and a piece of $5/32$" silver steel fitted. Drill this with the hob in place and everything secure.

Now replace in the machine with the index screw in place and tighten down the four $5/16$" clamping screws (the drawing

Grinding the front faces of the hob teeth using a toolpost grinder on the vertical slide.

41

Milling a 46 T. gear wheel, using the hob in the milling attachment.

paper has of course been removed). This makes a very solid fixture for milling. The cutter will clear easily towards the rear, but it is better to fit a limit stop to the forward feed of the cutter (or rather the rearward feed of the hob). By leaving the saddle free for the moment, the cutter can be eased into the groove and then the saddle locked. I drove the lathe in reverse; this is necessary with the fixture in this position. The depth is determined by taking trial cuts until the cutting edge at the bottom of the teeth is just touched up. Feed across, but stop just before the cutter fouls the tooth in front. Slacken off the four clamping screws, withdraw the indexing screw and with the tommy bar index to the next tooth, in the same row, and continue right around the hob. Slacken the saddle, move into the next row as before, clamp the saddle and continue round the next row. Do not forget the side clearance on the two outer teeth.

The backing-off fixture.

A hob set up ready for use.

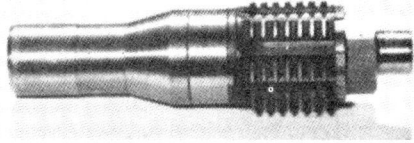

This will give a good side clearance and bottom clearance to the teeth and should remove the turning marks from both sides of all the teeth, right up to the cutting edge. The tops of the teeth are milled off at exactly the same setting using the 2½" × ³/₁₆" slotting cutter, one tooth at a time, in exactly the same way until there is just the faintest of witness marks left on the top cutting edge of each tooth. When you are satisfied that all is as it should be, with the tops of the teeth the same width or very slightly wider than the bottoms of the tooth spaces, the hob should be marked on one end with number punches **20 D.P. D 0.108"** and the burrs removed from the punching.

The hob is now hardened in the same manner as the cutter and the front face of the teeth are ground, using the same setting for gashing. I forgot to mention that when gashing, bring one of the cutting faces right in the centre of the tongue groove; subsequent hobs made, if gashed in the same place, will all come right for the backing-off angle. I have never had any distortion in the hole of a cutter I have hardened. Polish out with some fine 'wet and dry' paper wrapped around a piece of wood dowelling, with a slot down it to hold the paper. This will clean out the hole, leaving it exactly the same size as before hardening.

It is possible to grind the sides of the hob teeth, and this I thought of doing, using the same set-up as for backing-off but using a 2½" grinding wheel set up between two ball races in place of the milling spindle and driven from a high speed motor with a pulley on the shaft. But after trying out the hob, it cut so freely that I did not consider it necessary. If it becomes blunted on the

sides of the teeth, it can always be restored as described. The hob should however last very well between sharpenings, as cutting takes place on the whole of the 56 teeth as against the 10 or 12 teeth of an involute cutter. The first cut is rather heavy, but I had not the slightest trouble; I just fed a little more slowly, after the first cut it is no more effort than with an involute cutter. Do not forget that it is still just as necessary to set up the hob with the central tooth dead on the vertical centre-line as with an involute cutter, and feed through a second time after indexing half a tooth; or with the tooth on the centre line instead of the space move the hob along one half of the circular pitch. This finishes off and undercuts the bottom of the tooth space each side of the central tooth. Most of the cutting takes place on the near half of the hob if indexing is done away from the operator, so it is possible to even up any wear by indexing in a different direction for some of the gears being cut.

Perhaps I should give just the bare dimensions of the gear pitches for which I intend to make hobs, in case readers desire to make the same hobs.

D.P.	Circular pitch	Thickness of tooth on pitch line	Addendum	Whole depth
	in.	in.	in.	in.
16	0.1963	0.0982	0.0625	0.1348
20	0.1571	0.0785	0.050	0.1079
24	0.1309	0.0654	0.0417	0.0898
32	0.0982	0.0491	0.0312	0.0674
40	0.0785	0.0393	0.025	0.0539

CHAPTER 5

A Spherical Turning Tool

This spherical turning tool has several advantages over the more usual one that is bolted to the cross-slide of the lathe and usually uses a horizontal tool, generally adjustable in the toolholder to reduce the diameter turned.

It is fitted to the lathe with no more trouble than changing a tool. It will turn right down to the centre leaving no pip, if used with a half centre with the flat uppermost. It will turn to within $1/2$" of the chuck jaws. It will turn from 2" dia. to nothing with no difference in overhang of tool, indeed there is no overhang as the tool is end cutting and not subject to deflection or spring.

It is inherently accurate, simple to use and not difficult to make; materials to make it can generally be found in the scrap box.

Start by making the head part **A**. A piece of $2^1/_2$" b.m.s. shafting was parted off 1" full and faced off both ends. This was milled top and bottom to 2" length or it could be turned in the 4-jaw chuck. As I had a 60 deg. high speed cutter that I had made for another job, I used this for making the slideway after end milling to $1^1/_8$" wide × $^5/_{16}$" deep, the milling of the slideway being done in one cut. This could have been done by using a cutter only $^3/_4$" wide, in which case the slide piece **B** is milled first by bolting a piece of $1^1/_2$" × $^1/_2$" b.m.s. flat on a packing piece on the vertical slide and milling the two sides. This is then used as a gauge to mill the head **A** to fit to a fairly stiff push fit. No gib pieces are used as they are unnecessary if a good fit is made. In my case I had to mill part **B** afterwards which meant that I had to make and fit toolholder part **C** and also an

A view of the spherical turning tool in use.

Another view of the spherical turning tool in use.

additional piece of $5/8$" square steel at the other end and hold the two pieces in a vice and so mill the two angle faces (after a preliminary skim over the face) to fit the head. Make sure that the corners on the slide **B** are sufficiently removed to clear before fitting.

After head **A** is milled and slide **B** is milled to fit toolholder **C**, a piece of $5/8$" square b.m.s. $1^3/8$" long is held in the 4-jaw chuck with the centre $3/16$" from one edge but on the centre line, it is drilled $11/64$" for a depth of $3/4$" only and is then turned in this position to 30 deg. setting on the topslide to leave about $3/64$" of flat on the face. Do not drill right through at this stage. Reverse in the chuck, but pack out at front of one jaw and rear of the opposite jaw to tilt the piece about 5 deg. and central between the other two jaws. This is to give cutting clearance to the tool. Turn in this position to $1/2$" dia., leaving $13/16$" of the squared part and thread $1/2$" × 32 t. or 26 t.

Going back to slide **B**, the other side is reduced in width to $3/4$" leaving $3/8$" to stand proud of the head recess. Drill and tap $1/2$" × 32 or 26 at a position $5/16$" centre from the top edge; a second hole could also be drilled and tapped at the bottom in the same position (as I did) which can be used for a concave turning tool. Now fit tightly part **C** to slide **B**, filing the face a little to bring the tool hole to the top and square. Before final

fitting, clean thoroughly with petrol and flux with Easyflo on face and threads, then fit up and run Easyflo around the square face and the thread.

Now set up head **A** and slide **B** together in a machine vice in the drilling machine, with the top face of both pieces flush and drill down in the centre, half in head and half in slide, right through $5/16$". Open out in stages to $1/2$" for a depth of nearly $5/16$" and finish with a flat bottom to the hole using a $1/2$" D bit to $5/16$" depth. Reverse the pair in the machine vice, again bringing square with a piece of $5/16$" rod in the hole and bringing the two faces flush, open out as before to $1/2$" and finish once again with a $1/2$" D bit, leaving $1/4$" undrilled in the slide part. The shoulder left in the head part will be very little, due to bringing the faces flush each time. This shoulder should be removed with a round file in the head part only leaving a half shoulder in the slide. This is to fit the leadscrew.

The leadscrew could now be made in mild steel and to the dimensions shown to fit the slide. It is screw cut and threaded with dies to $1/4$" × 40 t. and graduations are made on the head (25) to indicate one thou per graduation.

Set up the head **A** in the 4-jaw chuck, milled face outward and $1/2$" off-centre vertically as drawing and turn a flat $11/16$" dia. to the depth of the bottom of the half round

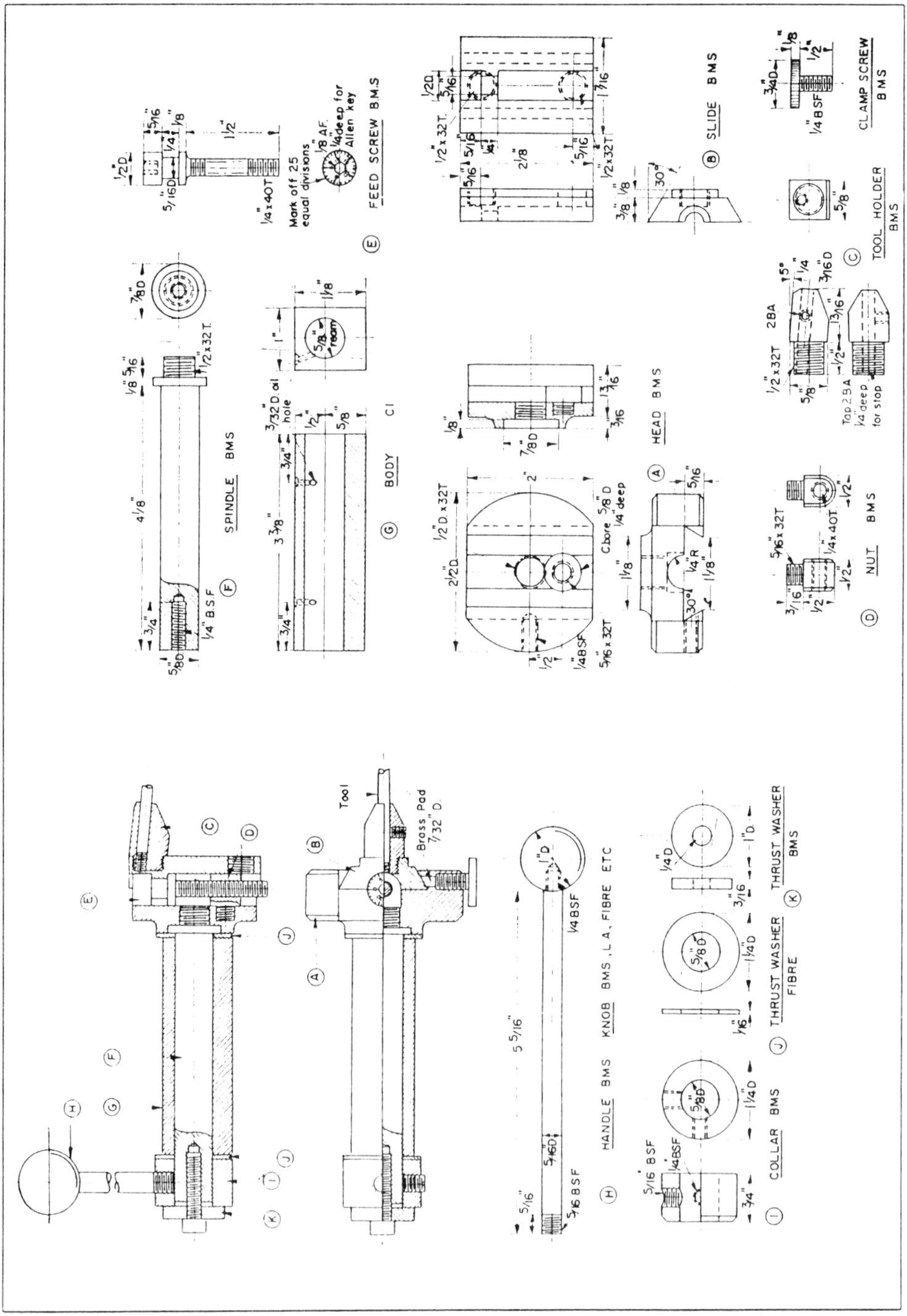

FEED SCREW B.M.S.

$1/2$ D

$5/16$" $1/8$"

$1/2$"

$5/16$ D

$1/4$" $1/4 \times 40$ T

$1/8$" A.F.
$1/8$" $1/4$ deep for
Allen key

Mark off 25
equal divisons

(E)

SLIDE B.M.S.

$1/2$ D

$3/16$"

$1/2 \times 32$ T.

$1/2$" $5/8$"

$5/16$"

$5/16$"

$2.1/8$"

$2.1/8$"

$1.1/16$"

$1/2 \times 32$ T.

$30°$

$3/8$" $1/8$"

(B)

CLAMP SCREW
B.M.S.

$1/8$"

$1/2$"

$3/4$ D.

$1/4$ B.S.F.

(C)

TOOL HOLDER
B.M.S.

$5/8$"

SPINDLE B.M.S.

$7/8$ D

$1/2 \times 32$ T

$1/8$" $5/16$"

$4.1/8$"

$3/4$"

$1/4$ B.S.F.

$5/8$ D

(F)

BODY C.I.

$1/8$"

1" $5/8$
ream

$3/32$ D. oil
hole

$3/4$"

$3.3/8$"

$5/8$"

$3/4$"

(G)

HEAD B.M.S.

$1/8$"

$7/8$ D

$1.3/4$"

$3/16$"

$3/16$"

(A)

$1/2$ D $\times 32$ T.

$2.1/2$ D

Cbore $5/8$ D
$1/4$ deep

$1/8$" R

$1/4$" R $5/16$"

$1/8$"

$30°$

$1/4$ B.S.F.

$5/16 \times 32$ T.

$1/2$"

$15°$ $1/4$"

$1/2 \times 32$ T 2 B.A.

$3/16$ D

$5/8$" $1/2$"

$3/16$"

Top 2 B.A.
$1/4$ deep
for stop

TOOL HOLDER
B.M.S.

NUT B.M.S.

$5/16 \times 32$ T

$1/2$"

$3/16$" $1/4 \times 40$ T

$1/2$" $1/2$"

(D)

HANDLE B.M.S.

$1/2$ D

$1/4$ B.S.F.

$5.5/16$"

$5/16$"

$5/16$ D

$5/16$ B.S.F.

(H)

KNOB B.M.S., L.A., FIBRE ETC

(H)

THRUST WASHER
B.M.S.

$1/4$ D

1"

$3/16$"

(K)

THRUST WASHER
FIBRE

$5/8$ D

$1/4$ D

$1/16$"

(J)

COLLAR B.M.S.

$5/8$ D

$1/4$ D

$1/4$"

$5/16$ B.S.F.

$1/4$ B.S.F.

$3/4$"

(I)

Tool

Brass Pad
$7/32$ D

(B) (J) (A)

(E) (C) (D)

(F) (G) (H) (I) (J) (K)

46

groove left by the $^1/_2$" drill. Drill and tap $^5/_{16}$"
× 32 for a depth of $^1/_4$". This is for the
leadscrew nut which is made to fit from a
piece of $^1/_2$" square b.m.s. The hole for the
leadscrew is best marked off and drilled with
the slide and head in place in the drilling
machine, marking off with a $^5/_{16}$" drill and
then drilling $^7/_{32}$" and tapping $^1/_4$" × 40 tpi.
This will ensure that the thread is in line.
The head is now replaced in the 4-jaw
chuck, this time with the back outside, and
run centrally. It is turned to $1^1/_4$" dia. for a
depth of $^3/_{16}$" drilled and tapped $^1/_2$" × 32 or
26 and counterbored $1^1/_8$" dia. to a depth of
$^1/_8$" for the shoulder of spindle, part **F**.

The main body part **G** should now be
made. I used a piece of $1^1/_2$" round cast-iron
milled or turned on four sides to $1^1/_8$" × 1"
and $3^1/_2$" long, but a simple pattern could be
made to save this work. Leave a little on the
bottom face for finishing truly flat. I hand
scraped this face to the surface plate.

This piece of cast-iron (steel is not
suitable) is now gripped in the toolholder
parallel with the lathe bed, using the barrel
of the tailstock or a parallel bar to bring it
truly parallel, the centre of the piece being
brought central with the lathe centres. It is
then centre drilled in place right through $^1/_4$"
dia. with the drill in the 3-jaw chuck and the
cross-slide locked. I used a piece of wood
between the tailstock and the piece for
feeding as then there is no twisting motion.
Finally it is drilled $^9/_{16}$" dia. I forgot to
mention that a boring bar made of a piece of

$^1/_2$" b.m.s. should be got ready first with a
tool from a short piece of $^1/_8$" round H.S.S.
held with a 4BA grub screw. This boring bar
is now passed through to bring the hole in
the main body to $^5/_8$" dia. Before removing
from the lathe, face off the end with a cutter
in the chuck and face off the reverse end
also to bring the length to $3^3/_8$". A couple of
oil holes are drilled and countersunk as
shown.

The spindle part **F** is turned from $^7/_8$"
b.m.s. $4^5/_8$" long and is made a nice sliding
fit in the body **G**; then it is screw cut $^1/_2$" to fit
the thread in the head **A**. The other end is
tapped $^1/_4$" BSF for the thrust screw, which
is adjusted to give a stiff movement to the
spindle. This spindle is now silver soldered
into the head and the Easyflo should run
nicely around the $^7/_8$" dia. shoulder to make
secure. Face this shoulder after brazing.

The toolholder **C** can now be finished
off. The $^{11}/_{64}$" hole is carried right through
and reamed $^3/_{16}$" to fit a piece of $^3/_{16}$ round
H.S.S. An Allen screw is fitted in the end of
this hole to form a stop for the tool and a
2BA Allen screw is fitted in the side
bearing on a small flat ground on the tool.
The tool is ground to 10 deg. *on the end
only* to give a 5 deg. top rake to the tool,
and should not be touched on any other
face. The tool is approximately $1^1/_4$" long.
A knurled screw $^3/_4$" dia. on the head and $^1/_4$"
BSF thread is made and fitted into the side
of the head as shown, this bears against a
$^1/_4$" piece of $^7/_{32}$" brass rod angled to 30

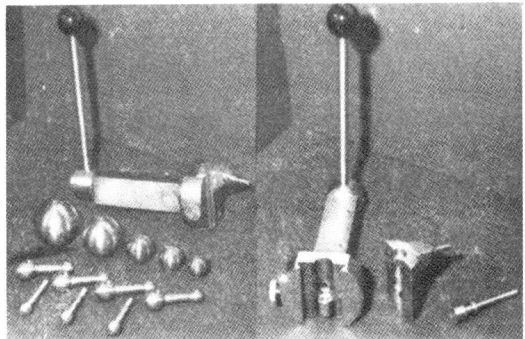

*Th finished tool together with some of
the work produced.*

deg. to clamp the slide during cutting.

Two fibre washers are made from $1/16$" sheet, $1 1/4$" dia. and $5/8$" bore to act as thrust washers, while the thrust collar part I and thrust washer part K are simple turning jobs. Make the thrust collar I a good push fit on the spindle and the $1/4$" Allen grub-screw should have the point flattened off. At right-angles, the $5/16$" BSF hole is tapped for the handle H which is a $5 13/16$" piece of $5/16$" b.m.s. rod. The handle can be your first turning job using the tool and can be turned in place on its own rod.

By fitting another toolholder in the bottom hole of the slide B and pointing the tool downwards, concave faces can be turned, and using the tool on a vertical slide instead of the toolpost, handles and hand wheels can be made.

In use, the tool edge is brought to the centre position and the cross-slide is locked. I use a half centre in the tailstock with the flat uppermost and I usually rough turn away the corners of the piece beforehand. The feed screw has a $1/8$" hexagon hole about $1/4$" deep, drilled $1/8$" and punched with a short piece of Allen key rod, for feeding, and after a preliminary touch over the high spots, a 25 thou cut or one turn is the usual feed. The slide is locked for each cut.

For turning diameters greater than 2" using the vertical slide, a specially angled tool could be made and the feed made not on the centre line of the lathe but in front of it. I have not tried this but it should work out all right.

The five steel balls illustrated are $1 3/4$", $1 1/2$", $7/8$" and $3/4$" dia., all in free cutting mild steel, turned at speeds of 425 for the two larger ones and 615 for the others. Machining time is between one half and one minute each. No tool marks were left on the work and the only finishing was a quick rub over with a bit of 220 'wet and dry' paper.

The four small clamp handles are only $1 1/2$" centres and the ends are $3/4$" and $1/2$" dia. turned from a piece of $3/4$" b.m.s.; the other three little handles are $1/2$" and are for a star wheel for an ejector for the rack feed tailstock.

CHAPTER 6

A Lathe Slotting Attachment

One of those necessary engineering operations usually done in the lathe by model and amateur engineers is slotting. This is usually done by mounting a boring tool under the toolpost with a flat-ended tool set on its side and the saddle is racked along the bed by the traverse handwheel. It is very distressing to most of us to feel the strain imposed on the saddle gears and to know the strain imposed on the slide ways by this operation. I have, in the past, been in the habit of pushing the saddle along by means of the tailstock which is a rack feed type with a pad centre pushing directly on the end of the tool.

By this means a lot of the strain is eliminated, but anything bigger than a $^1/_8$" wide keyway is not practical and even this takes a long time, also the tool will constantly move under the toolpost, upsetting the cross-slide reading.

I need to cut $^1/_4$" wide keyways in H.S. steel for milling cutters and many other jobs and generally have to cut them by hand. I have designed a special slotting attachment for the Super 7 lathe which, with some modification to the tee-slot hole positions, will suit the ML7 and other lathes. This attachment in conjunction with my dividing attachment will make it quite practical to produce such things as multiple splined holes to fit spline shafts, internal gears and internal shapes and cams.

The leverage obtained is about 20-1 and there is practically no strain on the slide ways and no wear on them. The work can be carried out without moving the part from the chuck and is inherently true. The permissible static load on the thrust bearings of the headstock is of the order of 3,000 lb. so there is no need to worry on this account. The hole for the ram is bored out in position on the cross-slide, so there is no longer any excuse for 'drunken' keyways.

The first thing to make is the pattern; this is quite simple. Start off with a piece of $^1/_2$" plywood $6^3/_4$" × $4^1/_{16}$" and a piece of $^1/_8$" plywood the same size. Cut the centre out of this $4^3/_8$" × $1^3/_4$" to form a recess to relieve the bottom face and glue it on. These sizes will allow $^1/_{16}$" on machined faces as the end facing the chuck and the operator is machined off $^1/_{16}$" and also off the bottom face. Allowing for this fix the $^7/_8$" discs which are $^1/_8$" thick in position as shown. If for a different lathe, position these discs, which are for the tee bolts, so that the centre of the ram will come within $^1/_2$" of the extreme 'in' position of the cross-slide. This $^1/_2$" is necessary when commencing to slot.

The centre pair of discs, while not strictly necessary, are to set the fixture one slot outwards for large diameter work without extending the cross-slide unduly. The rest of the pattern should require no explanation except to say that the core print for the ram should be $1^1/_4$" and about 1" long each end. Do not provide a core print on the operating shaft hole, as this would complicate the moulding and require a corebox. The casting will come back solid on this part, but is quite easily drilled and bored in place from the headstock of the lathe.

It is necessary to finish machine the ram before boring the casting as milling the rack teeth and keyway may cause a little

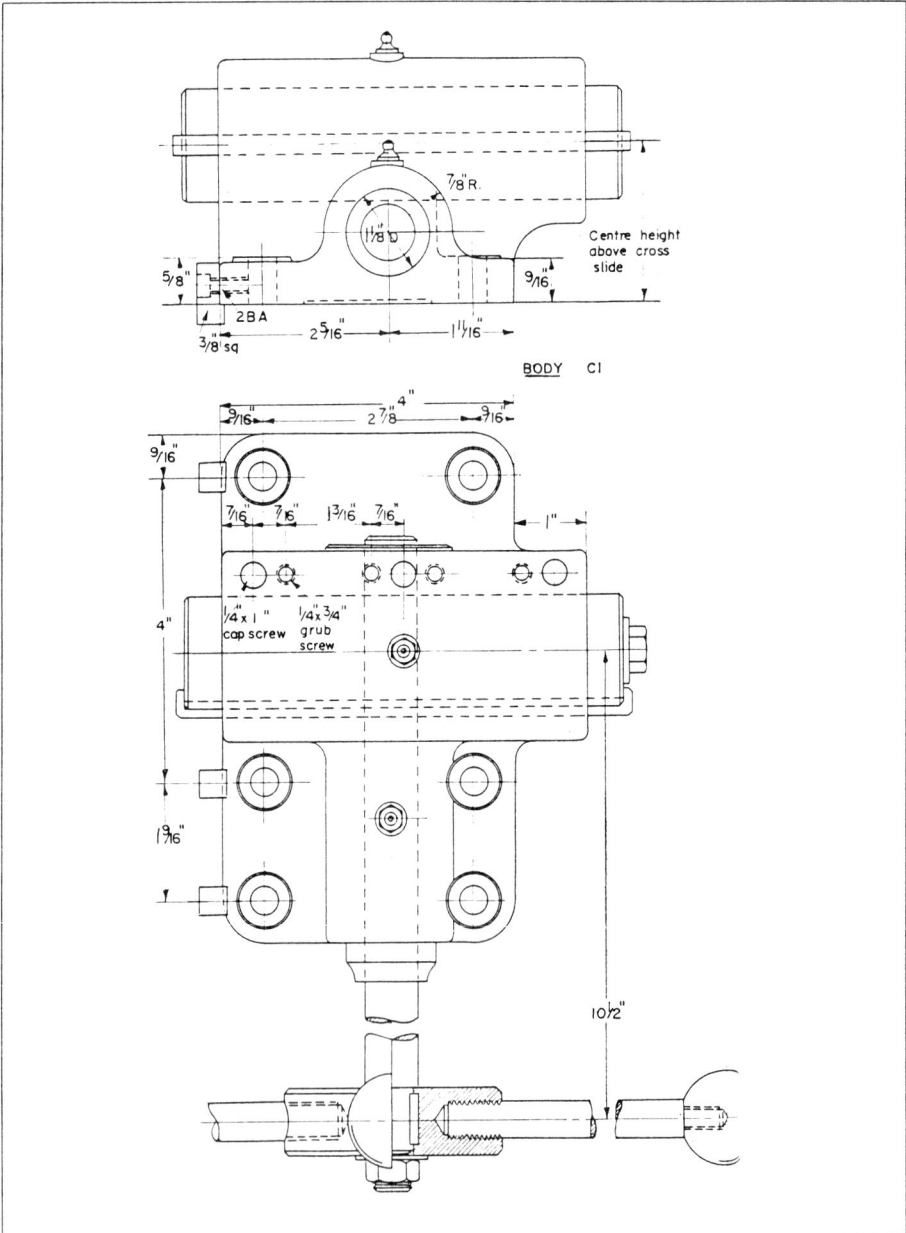

7/8" R.

1 1/8" D

Centre height above cross slide

5/8"

9/16"

2 BA

3/8" sq

2 5/16"

1 11/16"

BODY C1

4"

9/16"

2 7/8"

9/16"

9/16"

7/16"

7/16"

1 3/16"

7/16"

1"

4"

1/4"x 1" cap screw

1/4"x 3/4" grub screw

1 9/16"

10 1/2"

Milling the rack teeth using the 'Bormilathe' type elevatinghead. The vice is bolted to the quartering table.

distortion. A piece of ordinary $1^{1}/_{2}$" shafting is faced off each end to 6" long. A 1" hole for the various tools is bored in one end $2^{1}/_{8}$" long and a $^{3}/_{8}$" for a draw bolt is carried right through. Make a small bevel on each end at 30 deg. I set up this steel piece on a 1" spigot in the chuck and supported by the tailstock centre I was able to mill the keyway which is $^{1}/_{4}$" × $^{1}/_{8}$" and also the rack teeth with my milling attachment which was described in Chapter 1. The keyway could be end milled while held on the vertical slide and perhaps the easiest way to machine the rack teeth is in the slotter. Set a $^{1}/_{16}$" parting tool at $14^{1}/_{2}$ deg. angle and mill each side of the teeth and afterwards level off the bottoms of the spaces with the same tool. First however machine $^{1}/_{32}$" off the diameter and then mill to a depth of 0.135" spacing the teeth 0.1963" which is the pitch of the 16 D.P. rack. The best way to do this is to use a stop clamped to the slideway of the shaper with a prepared distance piece exactly 0.1963" thick inserted. Clamp the stop tightly and then remove the distance piece and you have the spacing for the next tooth. The rack teeth are 4" long, leaving 1" of 'plain' each end.

The key has ends bent over to prevent it working loose on its screws and also at the large hole end of the ram to form a key to hold the various slotting tools in the proper position. Merely heat the key to red and

forge over a sharp piece of steel held in the vice. The three screws to hold the key are 4BA Allen screws sunk flush with the surface. The key and the rack are at right angles in the position as shown. Now set up the ram on a true spigot in the chuck supported by the tailstock centre and test for straightness. If at all out, rake the smallest amount off the diameter to true it up. Make sure it is parallel, and polish while set up.

The casting can now be machined. First mark out and drill for the six tee slot bolts. Make sure the centre distances are correct and also that they are square with the long edge and with each other. Drill $^{25}/_{64}$". Make the tee-slot bolts and bolt the casting down in its proper place, packing under each bolt so that the casting is free from rock and is level and overhangs the front edge of the cross-slide $^{1}/_{8}$" right along. The front edge is now faced off $^{1}/_{16}$" using a facing cutter if you have one of at least $4^{1}/_{4}$" dia. or a flycutter to sweep about $4^{1}/_{2}$" diameter. This edge or face is now tested and trued up by scraping to a surface plate.

Now mount this true side face down on the cross-slide, letting the face of the base overhang the slide edge by $^{1}/_{8}$". Bolt with a long tee bolt through the cored hole and with two clamps. Using the same facing cutter or flycutter face off $^{1}/_{16}$" off the bottom, finish with a cut about 0.005" and then this

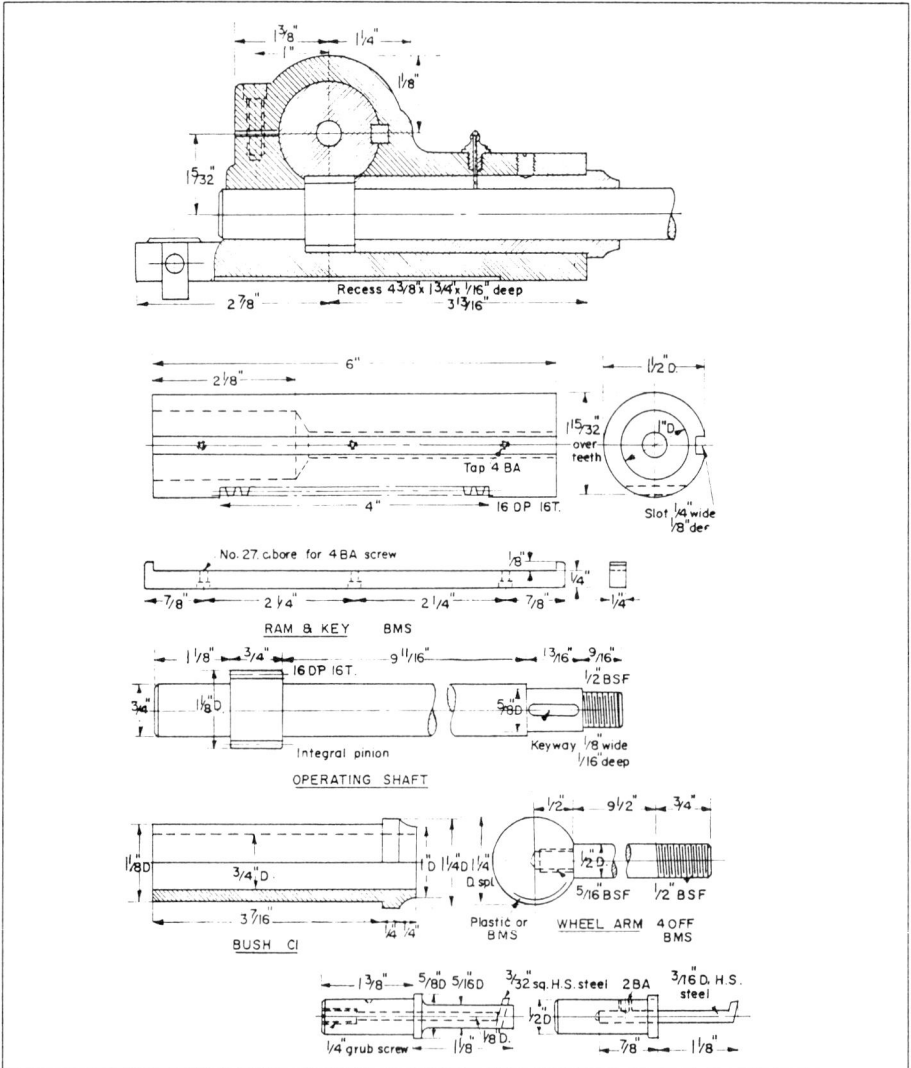

face must be hand scraped true to the surface plate. You will appreciate the large centre recess. The three stops along the finished edge must now be fitted. They are $^3/_8$" key steel $^7/_8$" long and are set in $^1/_{16}$" and fitted with a 2BA Allen cap screw set flush.

Make sure you mill or slot these $^1/_{16}$" deep grooves all the same depth. Spot face the tee bolt holes $^1/_{16}$" off.

Next bolt the casting down on the cross-slide in its proper position for boring for the ram. The boring bar can be at least 1" dia.

Boring for the operating shaft and bush.

with a $1/4$" dia. H.S.S. bit. Bore to a nice push fit on the ram, allowing all spring in the boring bar to run out. Now set a $1/8$" slotting tool on its side in the bar, lock the headstock spindle, and slot the keyway $1/8$" deep. Change the tool to a $3/16$" one and reslot and finally use a $1/4$" tool. Make sure the keyway is right on the centre-line and you can be thankful that this is about the second-to-last time you will need to abuse the lathe in this manner!

The centre distance between the ram and the operating shaft is $1^5/_{32}$" exactly, so the necessary packing pieces will have to be prepared of this thickness and set under the base with the casting at right angles. Allow the casting to overhang the cross-slide and after securely clamping, face off $1/16$" as before. This does not need to be checked for flatness but is now bored for the operation shaft bush. You will notice that this hole is offset $5/16$" from the centre. This

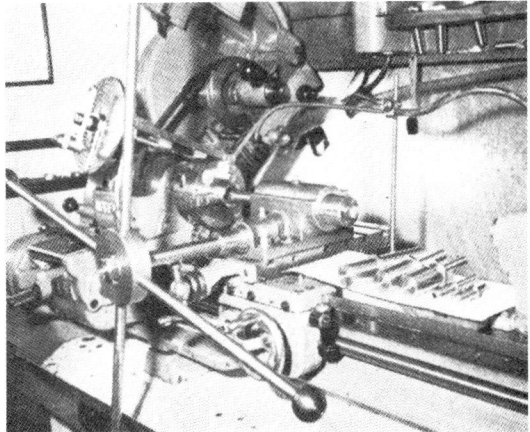

Slotting a keyway in a spacing collar.

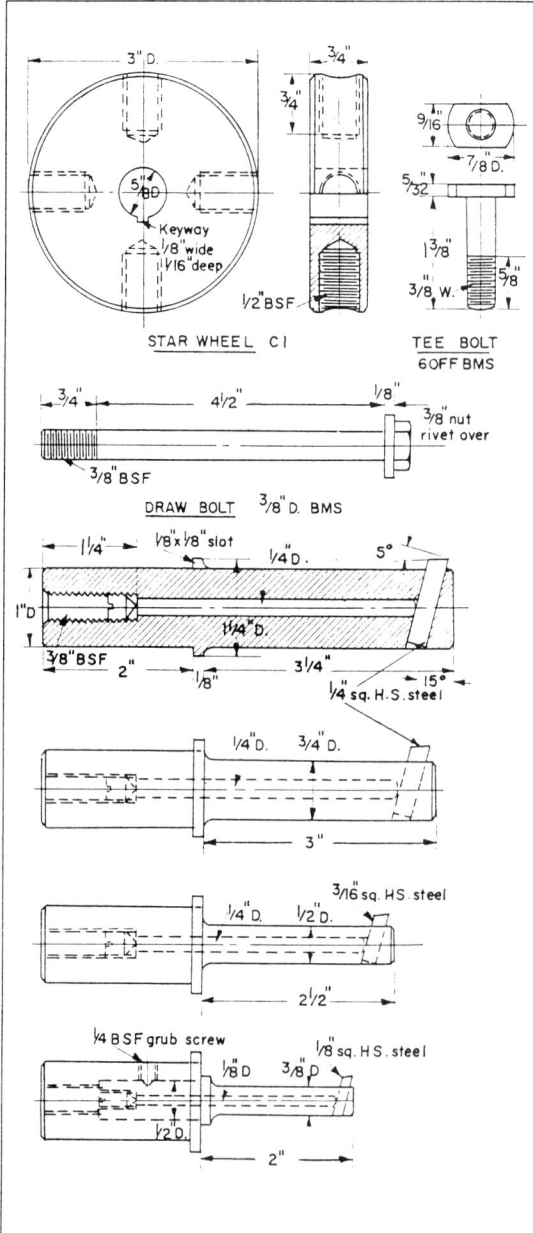

STAR WHEEL C I

TEE BOLT
6 OFF BMS

DRAW BOLT   3/8" D. BMS

is so that the operating shaft will leave the cross-slide handle index clear for reading the setting when slotting.

After clamping the cross-slide in position, centre drill with a large drill, and drill the size of the spigot on your drill to make a good start for the large drill. I have a 25/32" taper shank H.S. drill which is the largest No. 2 Morse made, and I used this in back gear to drill only to 4" depth, pushing along the saddle with the tailstock. This drill was followed with a 9/16" drill right through. This hole must now be bored to 1.125" for a depth of 4³/₁₆" leaving a flat face on the end of the hole. To do this I first bored the 9/16" remainder of the hole true to 3/4" using a boring bar between centres. Then, using a 3/4" dia. boring bar supported in this part and by the tailstock with the driving end held in the 3-jaw chuck, it was no problem to bore accurately to 1.125" or just slightly more so that the 16 D.P. pinion would slide into place.

The cast iron bush for the operating shaft should now be made. A piece of cored cast iron can be purchased from stock from most foundries, to finish turn to 1¹/₄" with a 1/2" cored hole. A 6" length will leave sufficient to hold in the 4-jaw chuck. Centre one end with a 30 deg. tool and turn a 2" length true to hold securely in the 4-jaw chuck.

With this turned part held in the 4-jaw and running reasonably truly, turn to clean up, bore to clean up and face, then proceed to finish turn the outside to a nice light tap in fit in the bored hole of the casting; the length of the turned part is 3⁷/₁₆". The boring is now done to 3/4" standard for a length of 4". A 3/4" *machine* reamer, taper shank spiral flute, will make an easy job of sizing accurately and parallel; leave only between 0.005" and 0.010" for reaming and a very slow speed in back gear but rush the reamer straight in as fast as possible, pause a second or two and then straight out again. The result is a bore than cannot be faulted. The bush is now parted off and the other end finished somewhat after the drawing, $1/2$" from the shoulder.

Star wheel hub

Before finishing with cast iron work so that the lathe can be cleaned for steel, the main casting should be slit with a $1/16$" or $3/32$" slitting saw $2 1/2$" dia. or larger, mounted on a stub arbor, and the work positioned above the saw on the vertical slide. Feed against the direction of rotation and bring the slit right on the centre-line of the bored hole. There is one other part in cast iron still to be made and that is the star wheel hub. This could be made of steel if you have a piece of 3" shafting $3/4$" thick or an odd bit of cast iron can be picked up from the foundry at the same time as the bush material.

It is turned all over to 3" dia., faced each side to $3/4$" thick and bored and reamed to $5/8$" and a $1/8$" key way slotted $1/16$" deep. This could with advantage be made $3/16$" and $3/32$" deep if you can manage it. Mount it on a stub mandrel and index four positions for the spokes, drilling through a piece of steel held under the toolpost with a hole about $3/16$" on the centre height. Drill into the cast iron $1/2$" or so in each of the four positions, in the centre of the casting. These holes can be enlarged in the drilling machine to $7/16$" for $1/2$" BSF tapping. With advantage they could also be drilled in the lathe with the drill in the chuck and the opposite hole supported in the tailstock centre. This will ensure they are truly radial. The depth of the holes is $3/4$" and the thread is finished with a plug tap.

Operating shaft

The operating shaft can now be made. A piece of $1 1/4$" shafting 13" long is centred each end and rough turned about $1/32$"

oversize all over, then finished as drawing. The integral pinion is 16 D.P. 16 teeth, pitch circle dia. is 1" and the outside dia. is 1.125", the depth of teeth is 0.135". I was able to mill the teeth and mill the feather keyway for the star wheel while still held in the chuck and on the centre, using my milling and dividing attachment. The cutter was one of my hob type cutters which is described in Chapter 4.

I am rather at a loss to recommend how this should be done if you do not have this or similar equipment. Perhaps a separate pinion could be made or you could obtain two 14 D.P. 14 tooth pinions from Bonds and reduce the shaft on the end to $1/2$" dia., fit a shallow key in the pinions and in the shaft and, leaving the shaft two or three thou oversize, shrink the pinions on. After turning to a total of $3/4$" thickness, a sleeve of $1/2$" bore and $3/4$" outside dia. and $1 1/8$" long is shrunk on against the pinions and pinned in place and finished on the outside to size. These pinions have a pitch circle dia. of 1", so the centre distances will be correct but the outside dia. is 1.143" so perhaps 0.018" could be turned off the outside without any harm. The rack on the ram will have to be a different pitch. The pitch of the 14 D.P. rack is 0.224in. as against 0.1963" for the 16 D.P. Bond's gears are however 20 deg. pressure angle, so the tool will have to be set to this angle and not $14 1/2$" deg. If the pinion has been reduced in dia., the depth can be the same, namely 0.135", but if not, the depth should be 0.154" and the operating shaft bush with its corresponding bored hole made to this size or slightly larger so that the shaft can be assembled.

The $5/8$" dia. end with the keyway is made a tight fit in the star wheel hub and the $1/2$" BSF thread screw-cut to fit a standard nut; a $3/4$" dia. x $3/32$" washer is also made to fit between the nut and the hub. The $3/4$" dia. parts should be a nice free fit in the bush and on the end spigot, and polished.

Star wheel arms

Four pieces of $1/2$" b.m.s. $10^3/4$" long are needed for the star wheel arms. The $1/2$" BSF thread $3/4$" long on one end to fit the star wheel hub should be screw-cut, as nothing looks worse than a wheel with spokes out of line. The $1/2$" long thread should fit the plastic ball threads if purchased or if steel balls are made they should be $5/16$" BSF. The steel balls should be blued in oil.

It remains now to make a set of slotting tools. I have drawn out a selection which should take care of all needs. The first three sizes, 1", $3/4$" and $1/2$", are all made from $1^1/4$" shafting, as also is the holding sleeve for the smaller sizes. There is a $1/4$" wide slot filed or milled in the $1/8$" wide shoulder which locates the tool on the correct horizontal centre-line. The tools are drilled and slotted at 15 deg. angle which makes for easy cutting and the end face angle should be 5 deg. back of the horizontal; the included angle therefore is 70 deg.

If the steel end is made a little large and longer until the square holes are slotted to fit the H.S. tools, gripping tightly in the vice and marking the tools will not matter, as the tools can be finish turned and faced after slotting. Slotting is done by drilling slightly oversize, filing with a square file to get a true start for the tool, and punching through with a square tool bit, ground square across the end. Get the hole dead on the centre line. The best way to do this is to set the tool up in the fixture on the lathe, clamping it at an angle on the cross-slide, and with a small flat filed on the end of each tool, the hole can be drilled; centre drill first of course. The push rods are of silver steel and the grip screws which press against the end are $3/8$" BSF $3/8$" long. The hole in the ends of the tools to accommodate these is $1^1/4$" long, as this hole also takes the draw-bolt.

Draw-bolt

The draw-bolt is threaded $3/8$" BSF for a length of $3/4$" (screw-cut) and a $3/8$" nut is fitted and riveted over on the other end; a $3/4$" dia. × $3/32$" washer is also fitted to take the thrust.

The smaller tools which fit the sleeve are merely held in the sleeve with a $1/4$" Allen grub screw; the screw provides the pressure to hold the tool through a piece of $1/8$" silver steel rod. The smallest tool is a piece of $3/16$" round H.S. tool steel held in a socket with a 2BA grub-screw, which fits in turn in the same sleeve as the other small tools. All the silver steel rods which hold the tools should have the ends which press against the tools rounded, and the grub-screws which provide pressure should have their points ground off flat.

To return a moment to the main casting, I find I have not mentioned a need for three 1" × $1/4$" Allen cap screws and the four $3/4$" Allen grub screws which clamp and hold apart the main bore for the ram. These provide precise adjustment for the fit of the ram and can all be tightened against each other so they will not work loose. A piece of oily felt should be inserted in the slot and a 2BA zerk nipple fitted in the centre of both this bore and the operating shaft bush. A $1/4$" Allen grub screw holds the shaft bush in place. The tee bolts, washers and nuts need no explanation.

This completes the construction of this fixture which readers will find makes slotting a pleasure and opens up possibilities of machining which would be impossible otherwise.

CHAPTER 7

Elevating Heads for the Lathe

The elevating heads with quartering table illustrated and described here, will be found most useful for many horizontal boring jobs. When the holes to be bored are too high above the crossslide or where holes are required at right angles or other angles and different heights (such as boring the cylinders and main and camshaft bearings for a small engine this can be done without moving the casting off the table. Horizontal milling can be done, the depth of cut adjusted by moving the heads, also face milling or end milling in different positions without unbolting, ensuring accuracy, and turning up to 15" dia. on the Myford without the limitation of the width of the gap.

I was inspired to design and make this fixture for the Myford by an illustration in *Model Engineer* of a 'Bormilathe' lathe described some time ago, but with my design there is no interference in any way with the normal function of the lathe, as the

General view of the Elevating Heads as set up for horizontal milling.

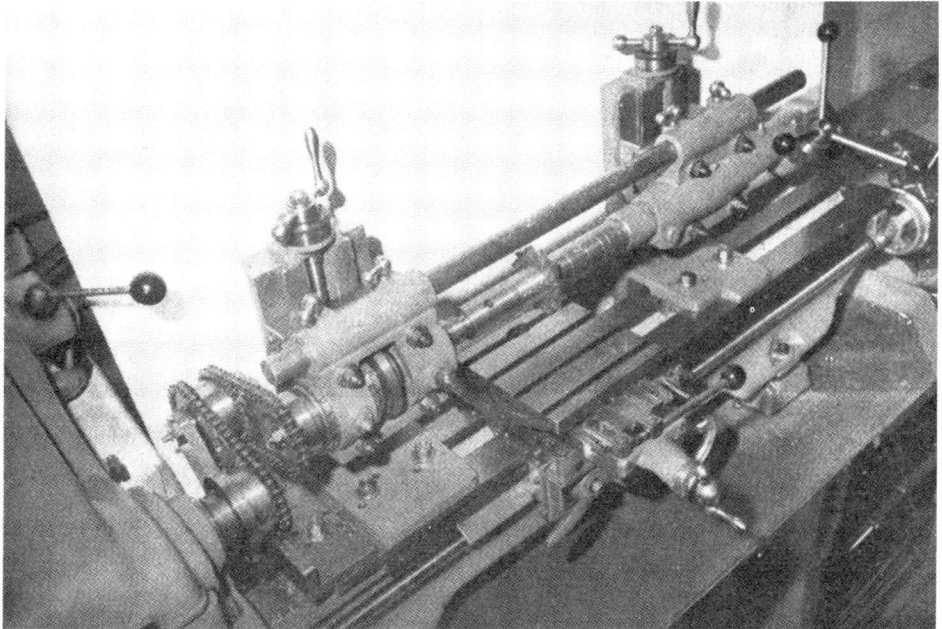

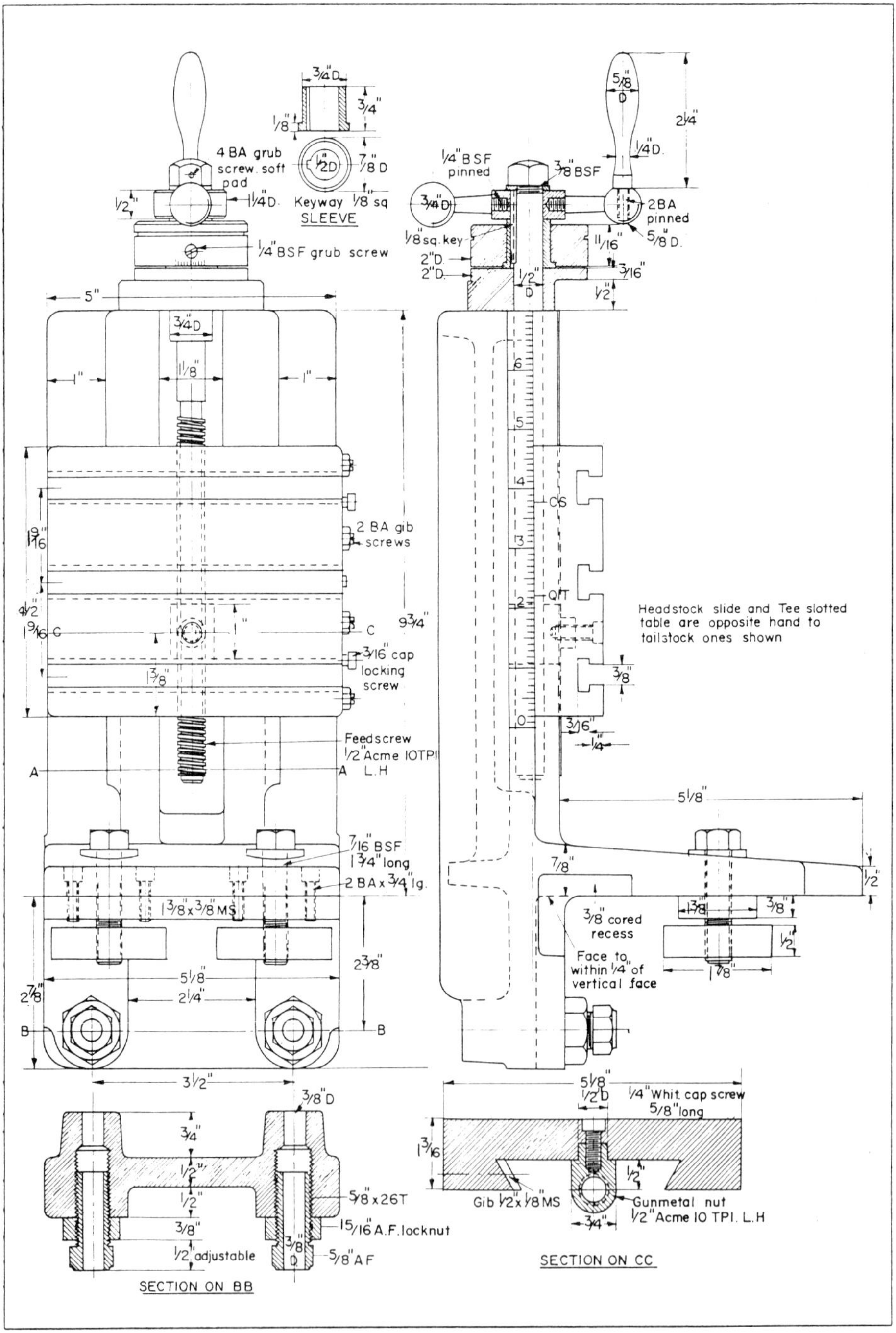

3/4"D
3/4"
1/8"
7/8"D
1/2"D
Keyway 1/8" sq
SLEEVE
4 BA grub screw. soft pad
1/2"
1 1/4"D.
1/4"BSF grub screw
5"
3/4"D
1"
1 1/8"
1"
2 BA gib screws
1 9/16"
4 1/2"
1 9/16
C
C
1"
3/16" cap locking screw
1 3/8"
9 3/4"
Feedscrew 1/2" Acme 10TPI L.H
A
A
7/16"BSF 1 3/4" long
2 BA x 3/4" lg.
1 3/8" x 3/8" MS
2 3/8"
5 1/8"
2 1/4"
2 7/8"
B
B
3 1/2"
5/8"
D
2 1/4"
1/4"D.
1/4"BSF pinned
3/8"BSF
3/4"D
2BA pinned
1/8"sq. key
11/16"
5/8"D.
2"D.
2"D.
1/2" D
3/16"
1/2"
CS
QT
Headstock slide and Tee slotted table are opposite hand to tailstock ones shown
3/8"
3/16"
1/4"
5 1/8"
7/8"
1/2"
3/8" cored recess
1 3/8"
3/8"
1/2"
Face to within 1/4" of vertical face
1 7/8"
3/8"D
3/4"
1/2"
1/2"
3/8"
1/2" adjustable
5/8" x 26T
15/16" A.F. locknut
3/8" D
5/8" AF
SECTION ON BB
5 1/8"
1/2"D
1/4" Whit. cap screw 5/8" long
1 3/16
1/2"
Gib 1/2" x 1/8" MS
Gunmetal nut 1/2" Acme 10 TPI. L.H
3/4"
SECTION ON CC

whole fixture can be removed in a few minutes. The two heads are set for the height required by the graduations in tenths on the sides of the slides and by the index collars in thousands; if the two settings are brought to the same figure, the stiffening support bar can be slid easily through the headstock and tailstock support bar hole, proving the heads are the same height exactly and in line.

The heads are very rigid in use being supported not only on the flat of the lathe bed, but also on the back machined surface of the bed, where the fixtures are rigidly bolted. Very rigid ribs run from these bolts right to the tops of the slides on the back of the castings and have proved to be quite firm, even with the heads well up on the slides. A 4" milling cutter can be accommodated under the support bar and the headstock and tailstock take all the normal fixtures of the lathe. Finally, by using a milling fixture, all the machining can be carried out on the lathe.

As usual, the first things to tackle are the patterns and core boxes. In order to support the main castings for milling in the lathe, I made the two main slide patterns as one, joined top-to-top with about $1^1/2$" of metal separating them, the one casting was then supported at each end on two specially made angle brackets and I was thus able to get enough traverse between these brackets to mill each slide at a time, the casting being reversed end for end for milling the second slide. The headstock slide has a plain foot but the tailstock slide has a cored recess and is cut away on the left front corner so that the saddle extension ways of the lathe can pass right along, allowing the saddle to come right up against the tailstock slide.

The two slides are identical, apart from this, and that the graduations on the vehicle edge of each slide are brought to the middle and the gib pieces of the tee-slotted tables are brought to the outside, making both the slides and tables right and left handed. Milling is done on the front faces and the two edges of the slides and all over on the tables. An allowance of $1/16$" on all machined surfaces was found sufficient and $1/8$" radius on all cored holes. The main pattern was made mainly of $1/2$" plywood, being straight and not subject to warping.

To simplify things, I have included a photograph of the patterns and core boxes where necessary, together with various other photographs, but I cannot illustrate the work progress by photographs as this fixture was made about three years before it occurred to me to publish anything. A left and right hand HS 60 deg. cutter and a left and right hand facing and radial cutter, $1^1/2$" dia. are required; these I made myself. Get the castings first and allow to weather outside while the necessary tools and jigs are being made.

I have included the pattern in the set up photograph to show how the milling is carried out. The slotted tables are milled with $1/2$" between the outer and inner faces, so that the slide faces are milled 0.010" shallow on the slides, as slotted tables ride on the outer and angular faces only. As the tables must finish up flush with the slide edges on the inner faces and are $1/8$" wider on the outer edges, it is necessary to make a pair of templates from $1/8$" mild steel or brass sheet which will fit each other accurately with the gib piece in place and to the dimensions shown, as it will be almost impossible to get this correct by measurement only.

The machine feed of the saddle can be used and the final cut should be about 0.005" to ensure accuracy and the minimum of hand scraping. All the cutting is taken to the left from the centre of the casting, hence the need for the right and left hand cutters. I had to use a 2" thick machined casting between the Myford vertical slide and my base which was faced and hand scraped on each side, in order to get sufficient height to mill the top edge. This simple 2" thick casting was $3^5/8$" x $3^1/8$", and long bolts were made to go right through.

The last inch of the slideways is cut away on the pattern as it is not possible to

machine right up to the face of the foot; this is cut away just sufficient to clear the cutter. Also the metal between the tops of the slides is cut away so that a free start can be made with the milling. I also milled the rear edges of the stiffening ribs along the back of the casting to micrometer from the front slideway face, to make it easy to set up for milling the foot and the top of the castings.

When finally the slides were milled to my satisfaction and after a sigh of relief and a cup of tea, I drilled a row of $1/2$" holes between the slides and broke it in half.

The method of fixing the casting to the two long angle-plates is to drill a $9/16$" hole in the centre, right under the foot and another near the top. Two $1/2$" bolts through the angle-plates, with a piece of heavy card interposed, held all securely.

The next job is to face the foot and top of each slide. This is packed up on the bed of the lathe between the tailstock and the saddle, bringing the milled edges parallel to the bed, using a surface gauge and dial indicator. The packing pieces of steel are placed close to and on each side of each bolt across the bed, to avoid distortion of either lathe bed or casting. A cutter is used as a facing tool, made from a piece of m.s. 3" dia. x $3/4$" thick with a single tungsten tipped tool pointing outwards at about 30 deg. This tool was mounted on my milling head on the vertical slide facing the tailstock, and it was not a great problem to mill to a nice finish and to within about $1/4$" of the back face. The pattern, by the way, was made with practically no draw at all on the two foot faces, all the draw on this part being on the top face to make the milling easier. The castings were then reversed and after

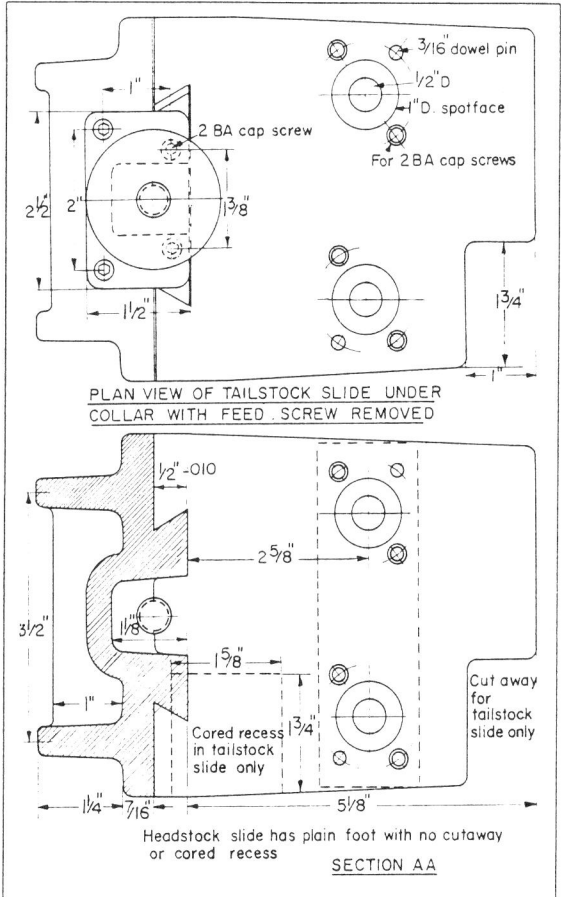

PLAN VIEW OF TAILSTOCK SLIDE UNDER COLLAR WITH FEED . SCREW REMOVED

Headstock slide has plain foot with no cutaway or cored recess

SECTION A A

removing most of the surplus metal by hand on this end face, it was also milled true, facing up the cross brace part which now forms the top face of the slide.

The next items to tackle are the two lead-screws from two pieces of $3/4$" b.m.s. $10^{1}/_2$" long. Turn one end to $1/2$" for $7^{1}/_2$" long and leave a $3/4$" collar $1/2$" long. At the other end, a $2^{1}/_2$" length is also turned to $1/2$" and the end of this is reduced and is screw-cut $3/8$" B.S.F. for $5/8$" long for the adjusting nut. The thread on the long end is $6^{1}/_2$" long, the last $1/4$" reduced to clear the screwcutting tool. The thread on this $6^1/_2$" length is $^1/_2$" Acme left-hand and both the turning and the threading should be done using the travelling steady. For the screw cutting I use the steady right over the top of the threading tool and produce very good threads. The threads are ten threads per inch.

Some time ago I made a left-hand and a right-hand Acme tap, some 6" long, from silver steel which I use to finish off the nuts. I also keep a gauge nut of each thread and so I was able to finish off these lead-screws to correct size. A keyway is cut in the short end with a $^1/_8$" end mill to $^1/_{16}$" deep at $^3/_4$" from the shoulder, finishing to $1^3/_4$" from the shoulder. This is to drive the sleeve which takes up end play when the adjusting nut is moved, leaving the index collar free to move.

The sleeve is $^3/_4$" long total and the indexing collar $^{11}/_{16}$" thick, so no side friction occurs on the index collar. The latter is 2" dia. bored $^3/_4$" to give a free fit on the sleeve with a $^3/_{32}$" deep recess for the $^7/_8$" × $^1/_8$" wide collar on the sleeve. The graduations are at the bottom (recess side), and there are 100 divisions. A 2BA Allen grub-screw is fitted on the zero mark which is locked in place when the settings are finalised.

The handle is made as drawing and now to complete this part of the fixture it is necessary to make the gunmetal nuts and the graduations on the slide and table. The nuts are turned to a $^1/_2$" spigot, tapped $^1/_4$" Whit. for holding screw, faced to 1" width and bored and screw-cut to fit screws. They are fitted so that the screws come central in the main casting and at $1^3/_8$" from the bottom edge of the tables. The top-plate for the lead-screw is an iron casting and is faced, bored and reamed $^1/_2$" for the screw, turned on the 2" dia., drilled for the four 2BA holding cap screws.

I have made the Cowell $^1/_2$" drilling machine with the milling table and I used these for the graduations on the edge of each slide. A pointed tool was set face down in a boring bar held in the chuck and the spindle of the machine clamped so it could not turn. The milling table was set up and the main slide casting set face down on this and parallel with the table. One turn on the graduated feed-screw of the milling table gave me each graduation and the depth was adjusted by providing blocks to three different thicknesses, which were interposed between the end of the tool and the milling table giving the depths for the graduations in $^1/_{10}$", $^1/_2$" and full inches. The zero mark was made at 7" from the top of the slide and the numbers were stamped with $^1/_8$" numbers. On the tables the mark CS is the zero mark for the cross-slide and the QT is the zero mark for the quartering table, which is exactly 1.500" thick so that the settings come right whether using the cross-slide or the quartering table. These marks cannot be made, however, until the whole fixture is completed. Meanwhile, you have two very useful milling tables on which long work can

The patterns, including those for the bracket (2 off) and small block for use while milling the main slides.

Tailstock and headstock set up for machining the rear face.

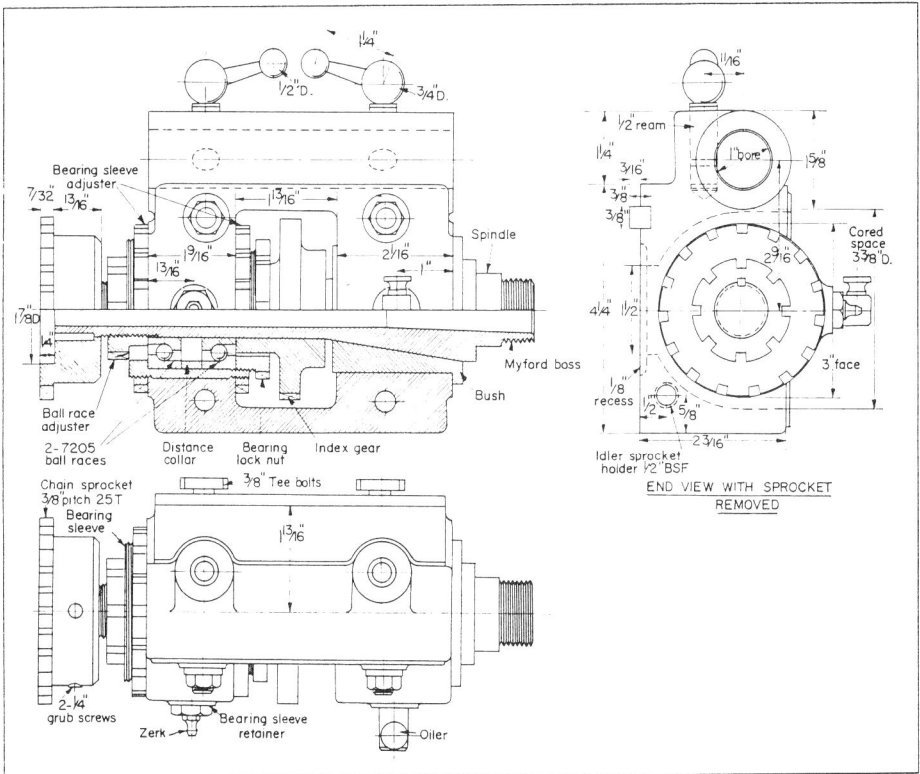

END VIEW WITH SPROCKET REMOVED

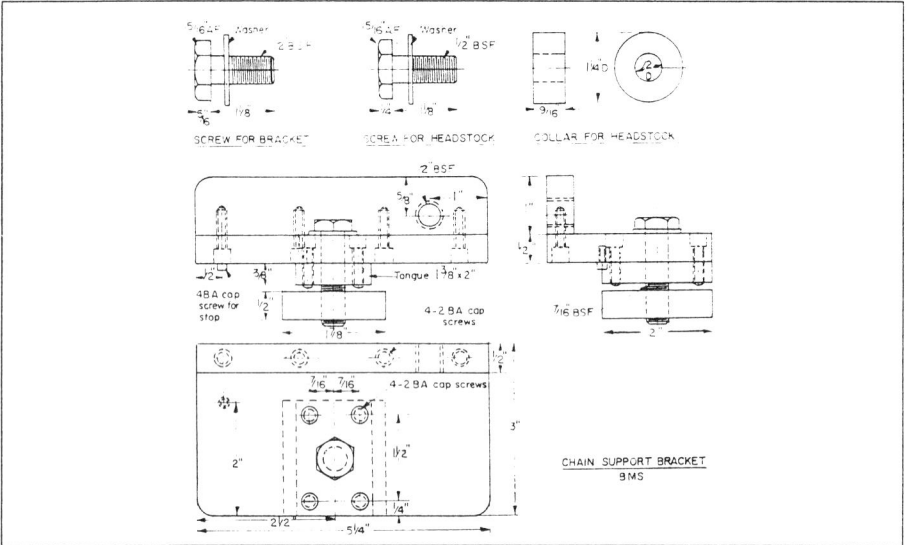

CHAIN SUPPORT BRACKET
9 MS

be bolted for milling, enabling the milling of machine slides, etc., not possible before, and also in conjunction with the milling attachment for facing and slotting the head and tailstock castings as will be explained.

I always have my castings sandblasted at the foundry, this saves a lot of destructive dust around the lathe and saves tool wear. Now set up the headstock casting on the cross-slide for taking a trueing cut over the bottom face. If four pieces of card are put under the rear face at each corner, and more used to remove any rock, the casting can be bolted down for facing. It is necessary to get the bottom edge or face true so that the rear face can be set up for facing and cutting the key-way. Make sure you set up with the two 3" faces in line as the whole alignment of the main bore will depend on this setting. Face with a large diameter facing cutter or fly-cutter to clean up.

This face is now set down on the cross-slide for facing the rear face of the headstock with the rear face towards the tailstock of the lathe; again bolt down true to the 3" dia. bosses and not to the cored holes. The amount to face off should leave the centre of these bosses at the same height as your centre height is above your cross-slide less exactly $1/4$". In my case this was $1^{13}/_{16}$". The milling attachment is now set up on the tailstock slide tee-slotted table so that the whole rear face of the casting can be faced up and the $3/8$" keyway milled. I used a $5/8$" end mill of the six cutting edge type for facing and this made an excellent job; one of the photographs shows this set up. Be careful that the key-way is at the same centre height as the raised bosses for the tee-bolts as the tee-bolts go through the centre of the keyway and misalignment would look very ugly. After facing, this face is hand scraped to a surface plate.

The two tee-slotted tables are machined next, being first faced up on the plain flat side in the four-jaw chuck. The pattern is $1^5/_{16}$" thick so $1/_{16}$" is faced off. The pattern itself has nicely rounded corners, so will clear the lathe bed. I always use tungsten carbide tipped tools made from $5/8$" key steel for cast iron and always obtain an excellent finish with a high degree of accuracy. After machining this face, the casting is clamped

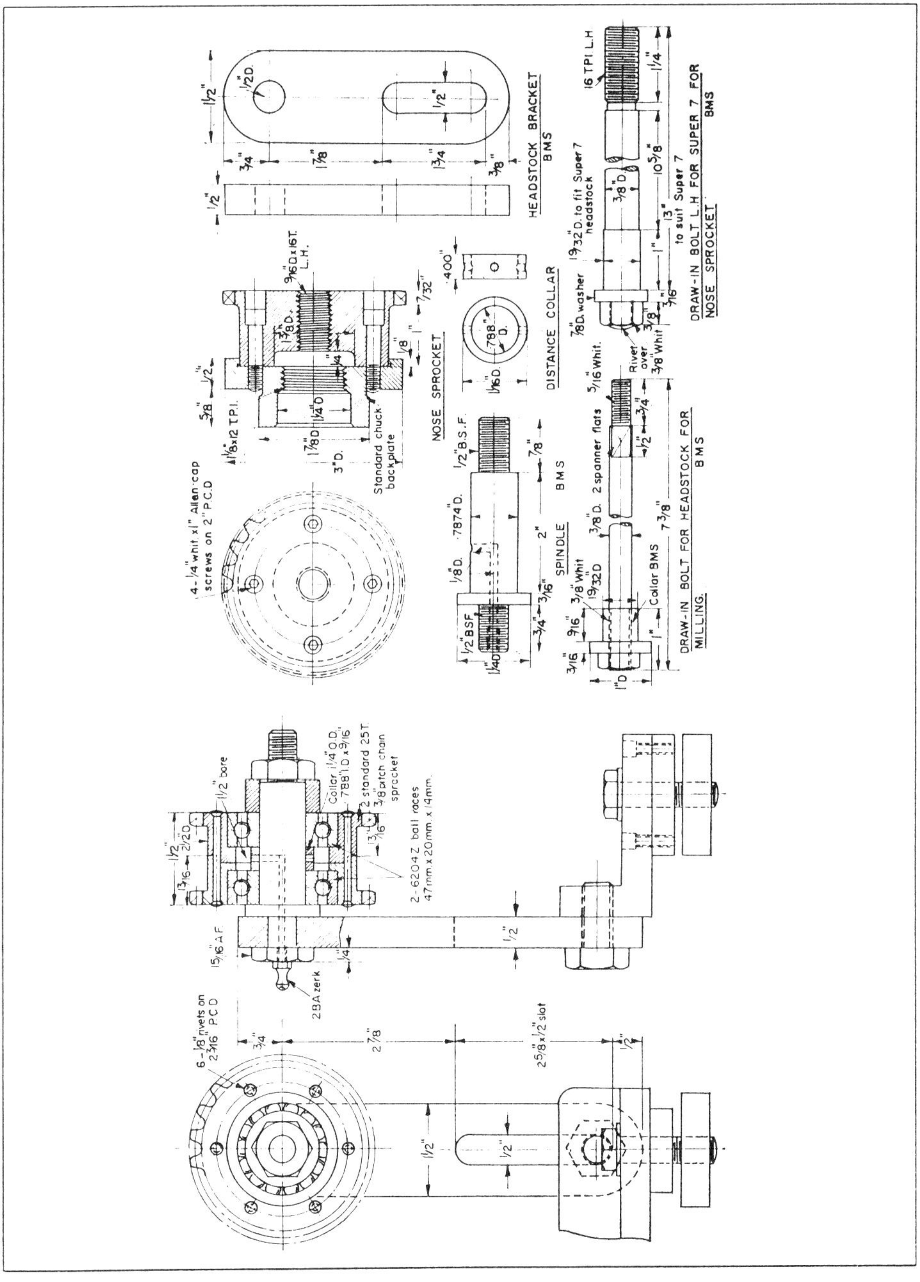
HEADSTOCK BRACKET
BMS
NOSE SPROCKET
DISTANCE COLLAR
SPINDLE
BMS
DRAW-IN BOLT L.H FOR SUPER 7 FOR
NOSE SPROCKET
BMS
DRAW-IN BOLT FOR HEADSTOCK FOR
MILLING.
BMS
16 TPI L.H
19/32 D. to fit Super 7 headstock
to suit Super 7
7/8 D. washer
Rivet over
3/8 Whit
5/16 Whit.
2 spanner flats
Collar BMS
4-1/4 whit x 1" Allen cap screws on 2" P.C.D
Standard chuck backplate
1 1/8 x 12 T.P.I.
9/16 D x 16T. L.H.
1/2 B.S.F.
.7874 D.
1/2 BSF
.788 D.
.400
1 1/2 bore
Collar 1 1/4 O.D. .7881.D x 9/16
2 standard 25T. 3/8 pitch chain sprocket
2-6204Z ball races 47mm x 20mm x 14mm
15/16 A F
2BA zerk
6-1/8 rivets on 2 3/16 P C D
2 5/8 x 1/2 slot

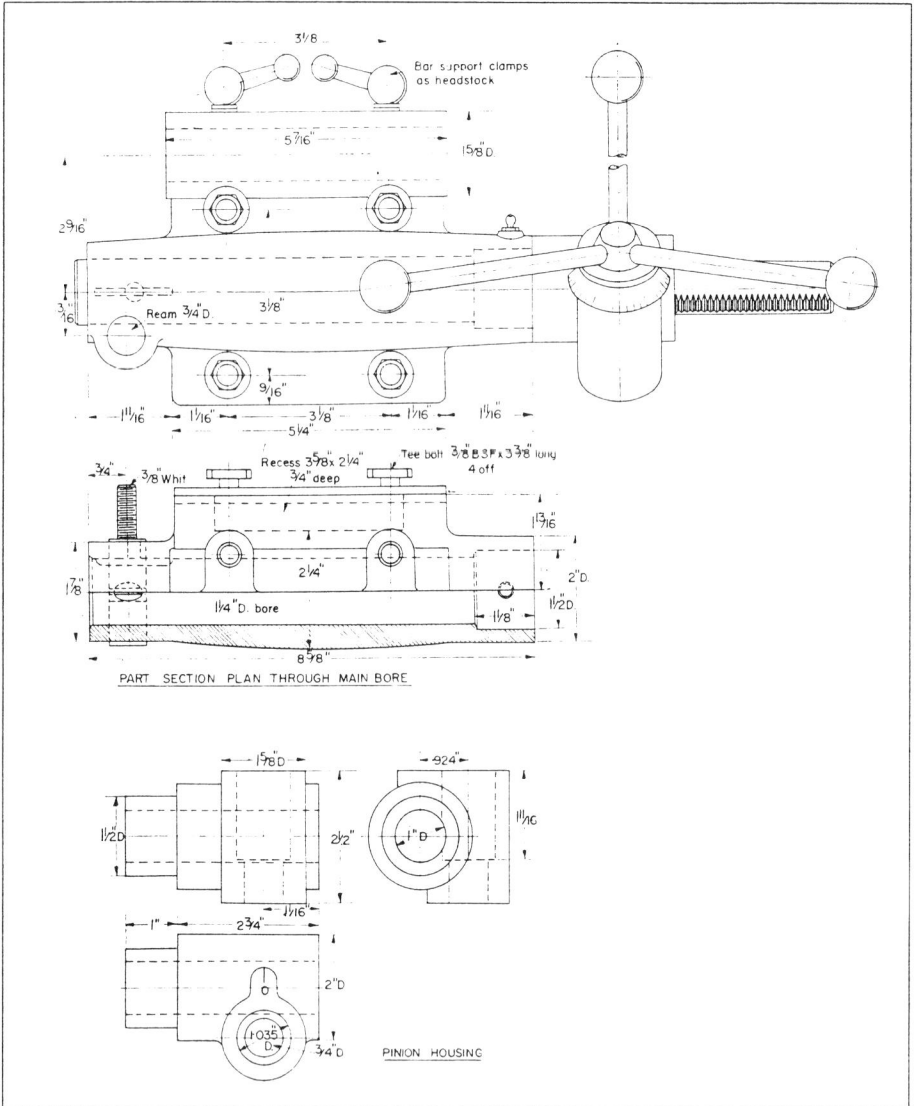

PART SECTION PLAN THROUGH MAIN BORE

PINION HOUSING

against the faceplate for machining the two $\frac{1}{2}$" high faces, the clamps being used on the lower inner face. This must be faced to $1\frac{3}{16}$", using the micrometer.

Next, each table is set up in just the same position on the faceplate but with the clamps on the two outer faces just machined. The milling attachment on the

1 1/16"

1/4" B.S.F.

3/8"D

3 3/4"

1/4"

3/8"

1"D bore

3 3/4"

1 1/4"

3/8"

5/16"

2 3/16"

3/8"

1/2"D

3/8" B.S.F.

15°

2"

2 — 3/16" steel balls

1"D plastic

13/16"

Compression spring

4 1/4"

3/4"D

Collar 1 1/4" O.D. 3/4" I.D. x 5/16"

4 B.A. Allen screw

1 7/8"

Keyway 1/8" x 1/8" full length of 1 1/4" D.

1 1/4" D

1"D

3/8" D.

No. 2 Morse

1"D.

3"

7 7/8"

6 7/8"

1/8" pitch

.086"

SPINDLE BMS

.985"

4 5/8"

At 45° to keyway

1/16"D

3/8"BSF

1 5/8"D.

1/8" Knurl

3/16"D

1/2"

1/2"

5/8"

1/8"

Recess 1/4" D. x 3/32"

3/32"

5/16"

48 graduations

1 7/32"

Oil hole

1/8" C.P

24 T

1·0345"D

·086" deep

1 1/8"

2 7/8"

2" D

3/4" D

1 1/4"D

1"D

1/4" D.

3/8"

1/2"

1/16"

KEY BMS
C HARDENED

1·034"D

3/8" Whit.
Washer

2 3/16"

1 5/8"

3/8"D.

Sleeve
3/4" O.D
3/8" I.D

1 — 1 1/16" turn off 1/16"
after turning

13/16"

1"

Tobin
bronze

2"D

Recess 1"D. x 3/32"

3/4"D.

SPINDLE CLAMP

JIG FOR BORING RACK
FEED CASTING BMS

2 circlips 25mm i.D.

1 3/4"D.

6202 Z
ball race

1 1/4"D

1/16" slot

1/4"

3/8" Allen

MILLING ARBOR SUPPORT BMS

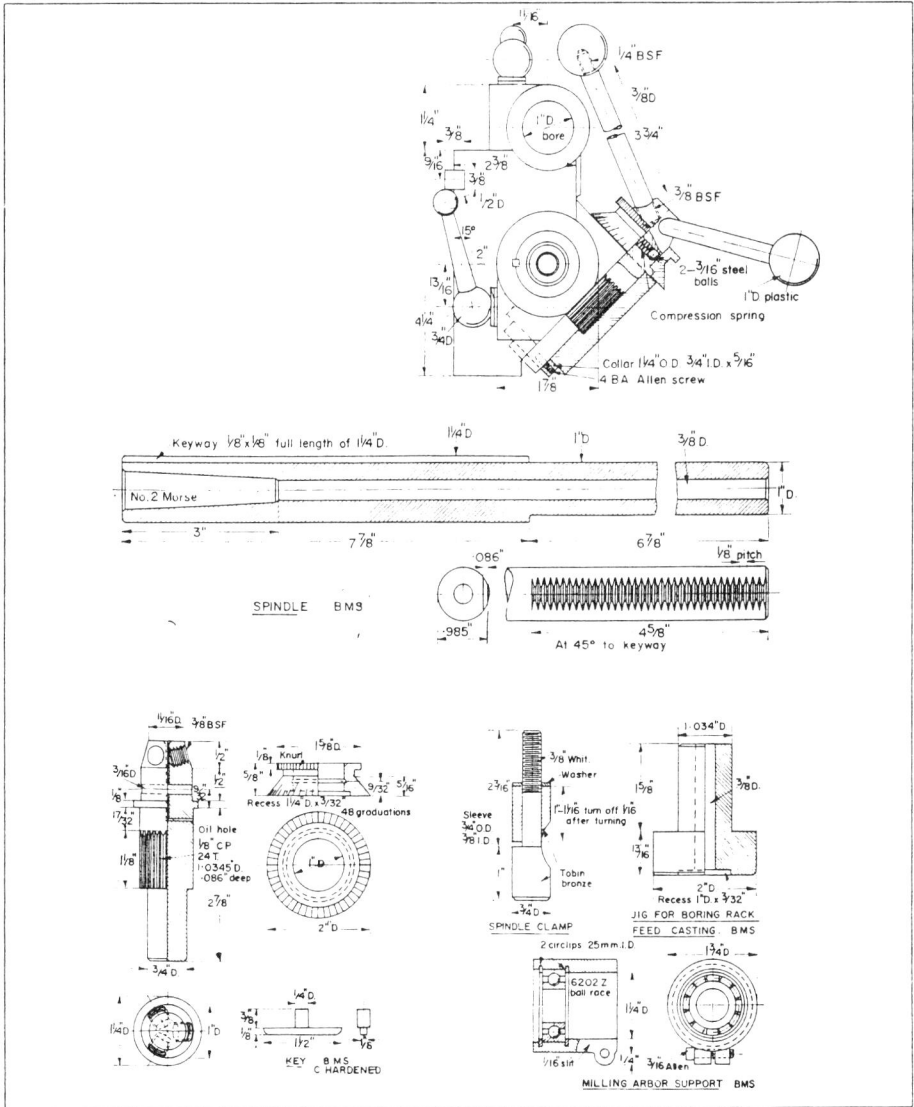

vertical slide facing the headstock is used and one of the 60 deg. cutters used for facing the slide is used for milling the inner face and the angular slides. The milling of the angle faces was not carried our to a sharp edge but the edges should be flat and about $3/64$" wide. To hold the headstock spindle firmly in place and in order to

Facing the foot of the main slides: the top was similarly machined.

position the casting in a truly horizontal position, I used my dividing head and was thus able to level the casting precisely for milling.

Each table was milled precisely to the gauges, and I was pleased to note that when finished and with the gib piece in place, I was able to slide the table over the

The tailstock elevating head complete.

slide. Reversing the table casting on the faceplate, I was able to mill the three tee slots at $1^9/_{16}$" centres, using first a $^3/_8$" endmill to $^7/_{16}$" depth followed by a home-made H.S. steel tee slot cutter $^9/_{16}$" dia. and $^3/_{16}$" wide. This is followed by the slow but necessary job of handscraping all the working surfaces to the faceplate and to each other. The tables are first handscraped to surface plate on both sides, and the angle face without the gib piece should be tested with a piece of ground gauge steel. Then the two tables are used as a surface plate in order to scrape the slide working faces, using artist's Prussian blue as marking.

My slides were just a little tighter as the slide went down, for some reason, but the amount was so small that I considered the accuracy excellent, in view of the unorthodox method of machining.

The tables are now set up on the cross-slide for milling the edges and a facing cutter is set up on the lathe spindle. The tables must be set absolutely square to both the tee slots and the slide ways, as this determines whether or not the headstock and tailstock are level. Great care is required over this and, after milling, the corners are rounded nicely to $^1/_8$" radius gauge and the edges polished. A piece of $^1/_2$" × $^1/_8$" b.m. steel is used for the gib piece, and there are four 2BA grub-screws with lock nuts and two $^3/_{16}$" Allen cap screws for locking the slides and also a $^1/_8$" dowel pin in the centre. The gib piece is held in place for drilling by clamping a piece of $^5/_{16}$" steel rod along it with two small clamps, and indentations only being made for all the screws. The dowel pin, however, goes right through, the edge of the hole being very lightly peened over with a small rivet punch to prevent the dowel riding right through, and also peened outside to prevent the dowel coming out.

The four adjustable sleeves with their locknuts should be made now: they are made from $^5/_8$" hex. steel with $1^1/_4$" of $^5/_8$" × 26 t.p.i. thread and the head part $^3/_8$" long.

The outer face is relieved as drawing and truly faced and a $^3/_8$" hole goes right through. The casting is drilled and tapped at $2^3/_8$" from the machined foot for these adjusters – note that there is a $^3/_8$" hole about $^1/_2$" long left for the clamping screw to pull against. The locknut for the adjusters is made from $^{15}/_{16}$" hex. steel and is faced off truly at $^3/_8$" thick. These adjusters are adjusted so that the headstock and tailstock are brought truly to the centre line of the lathe and until the centre gib piece which goes between the ways of the bed is dowelled in place, this cannot be done.

The two centre gib pieces are best made from $1^3/_8$" × $^3/_8$" b.m. steel flat - I had to use $1^1/_2$" × $^3/_8$" and make to a precise fit between the ways of the bed. Make also the clamping plates from 2" × $^1/_2$" b.m. steel, cut-off to $1^7/_8$" long, round all the edges nicely to prevent damage to the lathe bed and tap in the centre $^7/_{16}$" B.S.F. Four of these clamping plates are required. Make the four clamping bolts from $^5/_8$" hex. steel to fit the threads by screwcutting; the heads should be $^5/_{16}$" high.

In order to set the milling tables precisely to the correct distance from the centre line, it is necessary to turn a piece of cast iron in the four-jaw chuck to a diameter which will go over your cross-slide with a piece of $^1/_4$" steel flat interposed between the turning and the cross-slide. The reason for this is that the headstock and tailstock are bored on the cross-slide with a $^1/_4$" piece of steel packing underneath.

Using the same piece of steel (I used 1" × $^1/_4$" b.m. steel), turn the cast iron to this correct diameter. Now set up the headstock slide with the table clamped in place near the bottom and, using the adjusters, locate the tee-slotted table so that it just touches the turned piece of cast iron in the chuck. Then, using a dial indicator in the topslide of the lathe, bring the tee-slotted table parallel with the bed ways. Take care that the adjusters are hard up against the machined face on the back of the lathe and locked up in position; also clamp the slide down on the

Left and opposite: Two views of the main (driving) elevating head.

bed and check again to make sure the setting does not alter. With a scriber, mark under the foot along the lathe bed ways which will roughly locate the position of the centre gib piece. Fit the gib piece but drill the dowel holes in the foot only, not through the centre gib piece.

Set up the slide again with the centre gib piece in place, and if all correct tighten down the four 2BA screws which hold the centre gib piece. Remove the tee-slotted table and, with an electric drill, drill through for the two dowel pins which should then be fitted and then drill right through both the

foot and the centre gib piece. The two $1/2$" holes are for the holding $7/16$" bolts, right on the centre-line and at $2^7/8$" centres. Spot face the foot for the $7/8$" dia. × $3/32$" washers and then proceed in exactly the same way for the tailstock slide.

The headstock was not bored for the spindle or support bar until the tailstock casting was brought to the same state of machining, so that the two castings could be bored on the same settings.

First the four tee bolt holes were drilled at $3^1/2$" centres both ways, making sure the positions were square and as shown on the drawing. My pattern making was accurate as the bolts lined up beautifully in the centre of the raised bosses. The holes are $^{25}/_{64}$". The bosses were now spot faced with a home-made spot facing cutter and four tee bolts were made to fit the tee bolt slots in the cross-slide of the lathe. Two pieces of 1" × $^1/_4$" bright mild steel flat were cut off and drilled for packing up the casting $^1/_4$" for boring, these were placed length-ways with the cross-slide. A slot was cut in one of these holes in each piece from the side into the hole to accommodate a piece of $^1/_2$" × $^3/_8$" key steel, $2^{11}/_{16}$" long, fitted tightly into the keyway in the casting, between the tee bolt holes. This aligns the casting for boring. The casting is set up with the bar support cored hole away from the operator, so that the outer face of the left-hand cored hole, which is rather wide, can be faced off, which leaves only the inner face of this bearing hole to be faced with the boring bar and the smaller outer face of the right-hand cored hole for the bush.

I used a $1^1/_2$" dia. boring bar for the main hole and a $^5/_8$" followed by a $^7/_8$" boring bar for the bar support hole. All of my boring bars and milling arbors are supported at the tailstock end, not by the centre but by a ball race in a special holder clamped to the end of the tailstock barrel. A similar fitting is drawn among the parts of the tailstock and one was made for the tailstock. The headstock end is turned down to 1" dia. where the size is larger than 1" and is supported in the three-jaw chuck, the milling arbors being No. 2 taper shank, with draw-in bolt. The boring and facing tools in the boring bars are $^1/_4$" round for the larger bars and $^3/_{16}$" for the smaller, all of h.s. steel.

The centre distance between the two holes is $2^9/_{16}$", the graduations on the cross-slide being taken to get this accurately and

the bar support hole being bored to a push-in fit on 1" ground bar. The main hole which should finish up to a width of $5^7/_{16}$" between the outer faces, is bored $2^1/_8$" for the bush, faced outside to $2^1/_{16}$" long while the inner face does not need facing. The bore for the bearing sleeve is bored $2^3/_8$" and faced inside with the boring bar and outside with a facing cutter to $1^9/_{16}$" width. This facing should be carried out in 3" dia. to clear the two bearing sleeve adjusters.

Once again I had forgotten to mention that the two bar clamps in tobin bronze should be made and fitted before any boring is done. The bolt part, with its sleeve at $^{15}/_{16}$"

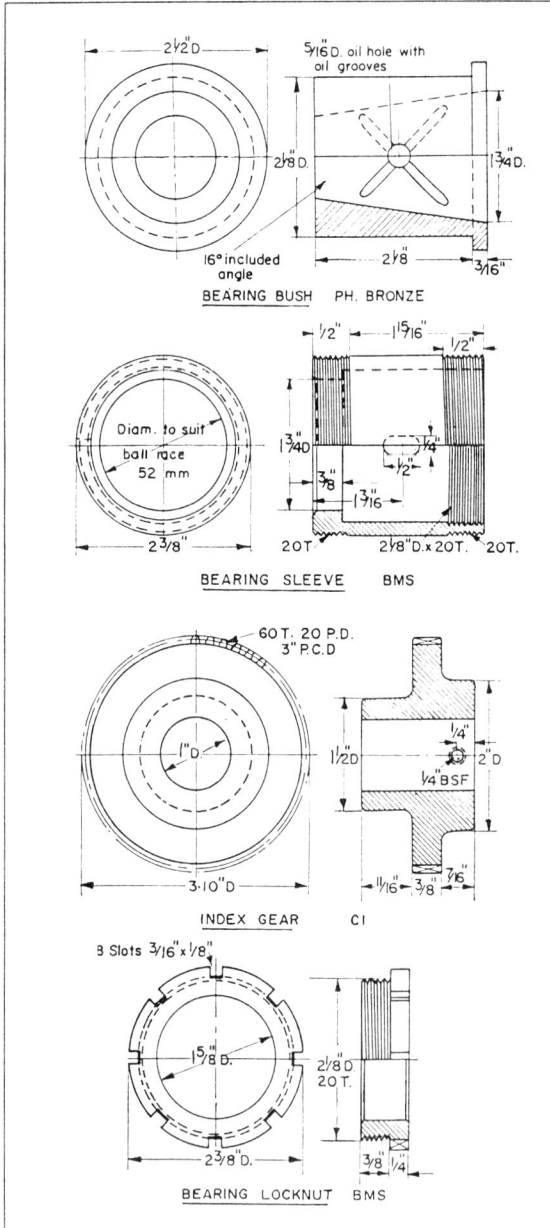

BEARING BUSH   PH. BRONZE

BEARING SLEEVE   BMS

INDEX GEAR   CI

BEARING LOCKNUT   BMS

long, should be clamped together and pushed into the $1/2$" reamed holes so that the joint comes central and secured in place with a 4BA Allen grub-screw while boring. This cuts the circular clamping faces to a perfect fit around the bar. After boring, the grub-screws are removed and $1/16$" is turned off the end of the sleeve part so that the two parts can clamp up. The centre of these two bar clamps is $11/16$" behind the centre of the bar support hole and at $31/8$" centres and it is reamed $13/8$" deep, $1/2$" dia.

After boring the casting, the two $7/16$" × 26 or B.S.F. holes for the oilers are drilled; the one for the zerk nipple holder, which also forms a retainer for the bearing sleeve, is at $13/16$" centre from the machined outer face and the one for the bush oiler is at 1" centre from its outer face. Any burrs inside should now be removed.

A tight-fitting gauge must be made that will just enter the hole for the bush, as the bush is turned all over at one setting and the outer diameter is turned next to the chuck and cannot be tried in the bored hole. The piece of phosphor bronze for this bush should be 3" long and a full $21/2$" dia. with a $7/8$" cored hole. Turn outside for $5/8$" long to clean up, and grip this part in the four-jaw chuck and then proceed to bore, turn and face. The outer shoulder should be $21/2$" dia. and $3/16$" wide. Turn up to the chuck jaws between three-quarter and one thou larger than your tight gauge for the outer diameter, reducing to

Some of the parts for the main elevating head.

your gauge the last $1/2$" or so at the end nearest the chuck to give a true start for pressing in the bush. The hole is bored 8 deg. setting on the topslide making it 16 deg. inclusive at the largest diameter of $1^3/_4$". The inner diameter should be quite a bit larger than 1" dia. at $2^5/_{16}$" long from the front face. Bevel the inner edge and face off at $2^5/_{16}$".

To fit this bush probably the easiest way is to put it for an hour in the freezer compartment of the domestic refrigerator and warm up the headstock until it is just too uncomfortable to hold. Then, using the balance of the bronze bar faced off each side laid on the bush and with a piece of end grain hard wood and a heavy hammer, it can easily be driven into place after first lightly oiling the bore of the casting. No other fixing is necessary. Drill a $5/_{16}$" oil hole through the bush and, with a small half-round chisel, cut a cross oil groove through the oil hole and remove all burrs.

The bearing sleeve is made from a 3" length of $2^1/_2$" bright mild steel and is also completely machined at one setting. The boring is done to fit the two angular contact ball races 7205B which are 52mm or 2.0472" outside dia. They should be a light tap-in fit. A 20 t.p.i. thread is screw cut for $1/_2$" long in the end of the bore for the bearing lock nut. Remove burrs from inside this thread so that the ball races are still free to enter, and turn the outside diameter to a tight push fit or light tap-in fit in the headstock bored hole. This sleeve must be capable of endwise movement under the action of the bearing sleeve adjusters, so must not be too tight a fit. A 20 t.p.i. thread is also cut on this diameter and also at the same tool settings on the end nearest the chuck. You will notice that the boring of this sleeve leaves a shoulder at $1^{15}/_{16}$" long from the outer face, the shoulder being $1^3/_4$" bore by $3/_8$" long, leaving the whole sleeve $2^5/_{16}$" long. The distance collar for spacing the bearings is a free fit in the bearing sleeve and is $3/_8$" wide and $1^3/_4$" bore with an oil groove around the centre on the outside

and with four $1/_{16}$" oil holes drilled right through to the inside. This and the bearing lock nut are in b.m.s. The bearing locknut has $3/_8$" of thread, a nice fit on the thread on the bearing sleeve and is $2^3/_8$" outside dia. on the shoulder. While still in the chuck, mark out with a pointed lathe tool the position of the eight slots for tightening up. All of these slots are milled at the same setting as will be explained later.

For the bearing sleeve adjusters a piece of 3" b.m.s., $1^1/_2$" long will be required. These two adjusters are turned $2^7/_8$" outside and are bored and screw cut to fit the bearing sleeve for the full width for the two plus the amount for parting off. Again mark off with a pointed tool the positions for the 15 adjusting slots on the outside and part off both adjusters to $1/_4$" wide.

The ball race adjuster, which adjusts the actual fit or clearances within the ball races and takes up wear and end play in them is made from a $1^1/_2$" piece of $1^3/_4$" b.m.s. and is best left until the shaft has been finished as it must be made a nice fit on the thread of the shaft. The shoulder on this adjuster is $3/_8$" long this time and is marked off for eight slots as before. A $1/_4$" Allen grub-screw is fitted in the centre of this shoulder, midway between two of the slots, and a soft pad is fitted under the grub-screw which must have the point ground off. The total length is $^{11}/_{16}$".

I would have liked to have made the spindle in case hardening steel and ground it all over after hardening, but it would have been necessary to grind the Morse taper and the thread on the nose and also possibly the thread for the bearing adjuster, so after due consideration I decided to make the spindle of 40 ton steel which has a certain amount of work hardening property. It was turned from an $8^1/_2$" length of 2" dia. steel, being first roughed out all over to within $1/_{16}$" of the finished sizes between centres and then carefully finished to sizes as drawing, both threads screw cut but the $1^1/_4$" dia. part on the nose left a little full, as this and the nose shoulder were finally finished in place, as also was the final finish on the Morse taper. The boring was done by setting up the fixed steady on the $1^3/_4$" dia. with the other end running true to the dial gauge in the 4-jaw chuck. A $^9/_{16}$" hole was drilled right through and bored to $^{19}/_{32}$" and the Morse taper bored, leaving a few thou for finishing. The final finishing was done on the $1^1/_4$" nose, nose shoulder and Morse taper while set up complete in the headstock on the headstock slide and driven by the chains in the normal way, the headstock being brought to the normal height of the lathe. By the way, the bearings should not be too tight a fit on the shaft as these must also be capable of movement for adjustment, particularly the outer one.

The index gear so far not described is 1" bore and 3.100" outer dia., for milling 60 teeth – 20 D.P. So far I have not milled these teeth, nor have I made the bracket for the worm or index plates, but these will be made as required, possibly very soon as I am now making a larger dia. clock face which will require a 12" indexed circle. The bearing sleeve retainer in $^5/_8$" A/F hex. steel prevents the sleeve from rotating while adjusting, by its $^1/_4$" round spigot fitting into a $^1/_4$" wide by $^1/_2$" long slot in the sleeve. The retainer is tapped for a 2BA zerk, for oiling the ball races. The plain bearing oiler is made from $^1/_2$" sq. key steel with a $^1/_4$" hole drilled for the round piece of felt kept into contact with the shaft by a spring; it is kept filled with oil and a little oil cap is made $^5/_{16}$" BSF with a knurled head as drawing.

The four tee bolts are made from $^7/_8$" mild steel and are $3^9/_{16}$" long × $^3/_8$" Whit., the head is $^3/_{16}$" thick and a $^3/_4$" × $^3/_{32}$" washer and a standard $^5/_8$" Whit. nut are required.

The slots in the two bearing sleeve adjusters, the bearing lock nut, and the ball race adjuster are all milled by clamping under the tool post of the lathe, packing up until the centre of the shoulder to be milled is at the lathe centre height, and the position of the slots judged by eye for the centre and squareness. No great accuracy is required for these slots and this method is quite satisfactory and quick. The cutter to be used is a $2^1/_2$" × $^3/_{16}$" slotting cutter on an arbor. The sprocket is a standard stock article – 25 teeth, $^5/_8$" pitch – and is the largest in b.m.s. stocked in this country, the larger ones being cast iron. Four of these spockets will be required and two $^3/_8$" roller chains of 24 links (48 rollers) with two coupling links also. The sprocket for the headstock is bored a tight fit on the $^7/_8$" dia. end and slotted for a $^1/_8$" key $^1/_{16}$" deep. A $^1/_4$" Allen screw is fitted to tighten down on this key, with the end slightly flattened, and another $^1/_4$" Allen grub-screw, at right angles to this, is rather deeply indented into the shaft.

On the headstock body drawing you will notice a $^1/_2$" BSF tapped hole, $^5/_8$" from the bottom face and $^1/_2$" from the rear face: this should be tapped about $^3/_4$" deep and is for one of the chain sprocket retaining bolts, described later.

The bearings are adjusted by first moving the bearing sleeve well to the right, so that the shaft is well clear of the tapered bush. The two bearings are then adjusted, by means of the ball-race adjuster, until all play is taken up in the bearing, but the ball-races are perfectly free, and the grub-screw is then tightened. Now, by moving the bearing sleeve to the left using the bearing sleeve adjusters, the tapered shaft is drawn into its tapered bearing until a good close fit is obtained: it should just be possible to feel the fit of this bearing. Pressure on the end of the shaft will not jam this bearing as pressure in either direction is taken up by the ball races. This is the same as in the Myford Super 7 lathe and is indeed practically the same size, only shorter and is, in my opinion, an excellent design. Keep the shaft adjusted until it is well run in.

The tailstock can now be machined. The bottom face will have been faced off at the same time as the headstock and also the drilling and spot facing of the tee bolt holes and the facing of the back face, all at the same settings. Also the support bar clamps will have been fitted and temporarily held with 4BA Allen screws and at the same time the clamp for the main spindle should be made and fitted in the same way with a screw at the back. The boring is done at the same settings as the headstock and at the same centres of course.

Before any of the boring is done, however, I would strongly advise you to make what I have described as a milling arbor support on the tailstock drawing. Make the fitting similarly but modified to fit your tailstock barrel, the same ball race, etc., being used. This is much better for all boring or milling than using a centre in the end of the shaft. Wear on the centre spoils the finish and reduces the bore size, expansion due to heat causes trouble and it is difficult to keep the centre adequately lubricated. The ball race overcomes all these troubles and your bores will be parallel and to a good finish. The main cored hole in the tailstock is 1" so I used a $\frac{7}{8}$" boring bar to start and I bored the right hand end first to $1\frac{1}{2}$" × $1\frac{1}{8}$" deep in order to reduce the length of boring bar necessary as the tailstock spindle of the lathe can go into this hole. The tool was put close to the end of the boring bar for boring this $1\frac{1}{2}$" dia. part. After roughing out the main bore to about $1\frac{3}{16}$" with the $\frac{7}{8}$" bar, a bar of $1\frac{1}{8}$" dia. was used for finishing the main bore to $1\frac{1}{4}$". This was actually bored 0.003" under $1\frac{1}{4}$" as I wanted to use standard $1\frac{1}{4}$" shafting which was about 0.002" under, and so I had to do a little finishing on the shaft to get a really good fit.

The ends of both bores are faced off to clean up and nicely bevelled inside and then the tailstock spindle is made. $14\frac{7}{8}$" of $1\frac{1}{4}$" shafting is cut off and at 7" from the chuck the fixed steady is bolted down. The four-jaw chuck trues up one end of the steel and it, of course, is running true at 7" from the chuck. The end is now centred, this procedure ensuring that the part of the steel which is not turned runs true. The centred end is now put in the chuck and the fixed steady brought to the end of the steel. The steel should be made to run true where the fixed steady has been before (not necessarily at the end close to the chuck) and the outer end is also faced and centred.

The $6\frac{7}{8}$" end is now turned between centres to 1" dia. and polished. Then, using the fixed steady again, but this time on the end of the steel and running true in the four-jaw chuck, it is bored right through from each end $\frac{3}{8}$" with a drill bronze-welded into a piece of $\frac{5}{16}$" b.m.s., the shank of the drill turned down to $\frac{1}{4}$". At this setting, while still running true, the Morse taper was bored and reamed, and the end slightly bevelled.

The rack is cut next, after first making a

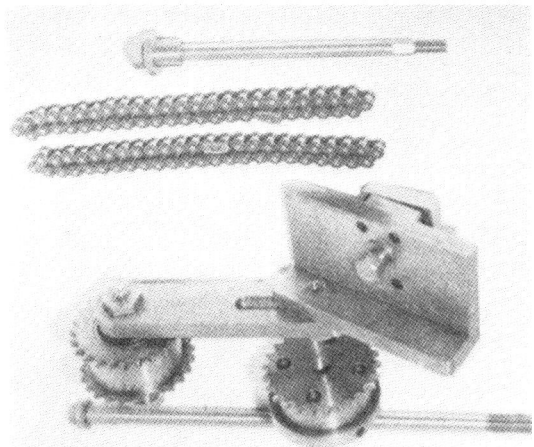

The chain drive mechanism.

The main slide pattern set-up for milling the main slideways.

flat along the 1" part for 4⁵/₈" long. The flat is 0.015" deep leaving the diameter across the steel at 0.985". I cut this flat and the rack in the hand shaper with a tool ground at 29 deg. inclusive, with a point $1/32$" wide. The pitch of the teeth is $1/8$". Before cutting the keyway which is at 45 deg. to the rack, finish off the $11/4$" part to a good push fit in the bore of the casting using a dead smooth file and 200 wet-and-dry paper with oil.

The keyway

The keyway is best left until the rack feed mechanism is completed, as the position of this housing can then be accurately determined and marked off through the $1/4$" reamed hole made for the key in the back of the headstock casting.

The rack feed casting is first faced and bored in the pinion hole, to 1.035" for 1¹¹/₁₆" deep, and the balance bored and reamed to $3/4$", the total width of this bore being 2¹/₂", this being done in the four-jaw chuck. Before boring the 1" hole for the spindle it is necessary to complete the rack pinion shaft. This is made from a 4¹/₄" piece of 1¹/₄" shafting, the pinion and pinion jounal being 1.0345" dia., the shoulder $1/8$" wide, the top part 1" dia. for $5/8$" long and the end at $1/2$" long being angled off at 15 deg. (30 deg. inclusive). The end shank is $3/4$" dia. to fit the

casting already bored, as also is the bore for the pinion. This pinion is $1/8$" circular pitch with 24 teeth, and as it is unlikely that a cutter will be available, a flycutter can be made. The thickness of the cutter on the pitch circle diameter, that is at 0.040" from the tip of the flycutter, should be exactly 0.0625" ($1/16$"). The total cutting depth of the cutter is 0.086" and the shape can be made the same as a 24 tooth change wheel, but note that the tool is very much thinner.

Milling the teeth

I milled the teeth with my milling attachment and the pinion should be rolled along the rack to remove any rough spots. To get the exact centre distance for the two bored holes in the rack feed casting, hold them together in close and tight mesh and measure with the micrometer over the shaft and the pinion. The centre distance will be one half the pinion diameter plus one half the shaft diameter plus clearance. For instance, if your measurement over the pinion and shaft held tightly together is 1.940" as it should be, though a small variation does not matter, add the clearance – say 0.001"–0.0015", that is 1.9412". Half the pinion diameter over the outside of the teeth should be 0.5172" – plus half the shaft diameter at 0.500" is 1.0172", which subtracted from the total width of 1.9412", leaves 0.024" as the centre distance of the two bored holes.

Boring the 1" hole in the casting at exactly the correct centre distance requires a jig fixed to the cross-slide of the lathe. This is made as the drawing from 2" m.s. and is turned a close fit to the 1.035" bored hole, and with a $^3/_8$" hole for a tee bolt. This tee bolt should be $4^1/_2$" long with about 2" of thread. This is bolted down with a nut approximately in the centre of the cross-slide and must be brought to the exact centre line of the lathe so that the cross graduation can be set to zero. This is done by bringing the turned diameter of the jig against a piece of steel turned truly to say $^1/_4$" dia. Bring the jig against this turned piece and take a reading; then against the turned piece at the rear and take another. Half this distance, allowing for half of the backlash, will give the centre position of the jig; then, being careful of the backlash, move the cross-slide the required centre distance and lock the slide. The casting is now bolted down with a washer and another nut in line with the centres. The height of the flange at the bottom of the jig is marked as $^{13}/_{16}$", which will bring the casting on the vertical centre height if the lathe is $2^1/_{16}$" centre height above the cross-slide. The 1" hole is now bored with a boring bar to a nice slide fit on the spindle.

At $^3/_{16}$" above the $^1/_8$" flange of the pinion shaft, a $^3/_{16}$" hole is drilled to accommodate a stiff compression spring and two $^3/_{16}$" steel balls, and in the index flange (made from 2" dia. m.s.) at $^9/_{32}$" from the bottom face, a very small groove is bored inside the 1" bored hole for these steel balls to run in. This prevents the index flange from moving upwards. It cannot move downwards or rub on the casting (which might alter the reading) because of the $^1/_8$" flange on the shaft and the recess in the index flange which is only $^3/_{32}$" deep. These steel balls and the spring give frictional position to the index flange. A b.m.s. collar with a 4BA grub-screw holds the pinion shaft in

position. The three tapped holes for the star wheel arms are tapped $^3/_8$" BSF and are at 15 deg. to the horizontal.

The tailstock spindle with the rack-feed mechanism can now be set up in the tailstock and a piece of 1" steel put into place in the bar support hole. A position at approx. 45 deg. will be found where the star wheel of the rack-feed clears the bar support, and the plastic knob nicely clears the tailstock body by about $^1/_2$". In this position, mark out the position for the keyway in the spindle through the $^1/_4$" reamed hole for the key. The keyway can now be milled in the spindle through this marking and should be $^1/_8$" wide by $^1/_8$" deep. The key is made from b.m.s. and is case hardened.

The rack-feed body is held in position in the tailstock body by three $^1/_4$" Allen screws. These screws are turned to $^3/_{16}$" parallel on the end and fit into three $^3/_{16}$" drilled holes; pointed grubscrews would close down the casting on the shaft and distort the bore. Four tee bolts $3^3/_8$" long complete the tailstock.

Chain Drive

The chain drive mechanism should be made next. The part described as the chain support bracket I made from b.m.s. plate 3" × $^1/_2$", with a piece of 1" × $^1/_2$" screwed along one edge to form an angle bracket. This could be made quite easily in cast iron from a simple pattern. The bottom face must be true to surface plate and the tongue piece of $1^3/_8$" × $^3/_8$" × 2" long, screwed along the bottom in the position shown. The clamp plate, $1^7/_8$" wide by $1^1/_2$" long, is clamped up by a similar screw to the heads $^7/_{16}$" BSF, $1^5/_8$" long. The $^1/_2$" BSF thread in the upright part supports the main chain support arm to which the sprockets are fixed. This arm is a piece of bright m.s. $6^3/_4$" long of $1^1/_2$" × $^1/_2$", slotted $^1/_2$" wide × $2^5/_8$" long, through which the $^1/_2$" BSF hex. screw screws into the support bracket. The two standard, 25-tooth, $^3/_8$" pitch, chain sprockets run on two 6204Z ball races, which are 47-20-14mm ball races with one shield. The shields go outside in the sprockets.

These two sprockets, which are stepped and must run true, are riveted together with $6^{1}/_{8}$" steel rivets made from $^{1}/_{8}$" round m.s. I first set up the sprockets in the four-jaw chuck with teeth facing outside and got the $2^{1}/_{2}$" dia. body of the sprocket to run dead true to the dial gauge. It was then bored out to 1}" and bored again to a light tap fit for the ball race at 1.8504" for a depth of 0.5512" leaving a square corner. Then, gripping with the internal jaws of the four-jaw chuck and using the ball bearing cone centre described in a previous chapter to line it up, it was held tightly against the jaws and set to run true again to the dial gauge. The shoulder on a diameter of $2^{3}/_{16}$" was turned male and female to a good fit, both sprockets being $^{13}/_{16}$" wide overall. For the shaft you will notice there is a distance collar: this must be two or three thou wider than the combined thickness of the internal shoulder left in the sprockets when they are tightly pulled together, or damage and very short life will result to the ball races. The shaft is made a light press fit for the ball races and both threads should be screw cut. A zerk nipple and grease holes must be drilled and an internal groove with three or four small grease holes drilled radially through the collar of the shaft.

The headstock bracket is also made from $1^{1}/_{2}$" × $^{1}/_{2}$" bright mild steel and is drilled $^{1}/_{2}$" for the headstock screw and a collar $^{9}/_{16}$" wide goes between this bracket and the headstock. The headstock should have been spot faced lightly to bring this bracket square when pulled tight. The other end of the bracket is slotted $^{1}/_{2}$" for $1^{3}/_{4}$" long to within $^{3}/_{8}$" of the other end. The easiest way to slot these heavy brackets is to drill $^{1}/_{2}$" at each end, and a row of $^{7}/_{16}$" holes between, remove most of the metal roughly with the hacksaw and set up on the vertical slide with a $^{1}/_{2}$" end mill in the chuck. A piece of say $^{1}/_{4}$" plywood between the vertical slide and the steel bar will protect the slide.

The nose sprocket is now made. I

purchased a standard chuck backplate, turned it to 3" dia., turned it true on the face and made a stepped recess to fit the body of the fourth sprocket; the corner was undercut and carefully fitted in the same manner as a chuck would be. A recess on the inside of the sprocket about $^{1}/_{4}$" wide and $1^{3}/_{8}$" dia. was turned and four $^{1}/_{4}$" Allen cap screws fitted on 2" centres. Before turning the backplate however, drill a $^{1}/_{4}$" tommy bar hole in the edge of the flange or difficulty will be experienced in removing the backplate from the lathe mandrel. With the sprocket set up on the backplate, bore through the sprocket $^{1}/_{2}$" and screw cut LEFT-HAND a 16 t.p.i. thread for a draw bolt. This draw bolt is necessary as most of the milling will be done with the lathe running in reverse and a right-hand thread will definitely not hold the sprocket on the nose. The draw bolt for the Super 7 lathe is 13" long up to the thick washer and made to the drawing. I have also drawn a draw bolt for a milling arbor for the headstock of this fixture which must be screwed into the arbor before inserting in the headstock mandrel; as it cannot be put through from the outside, two flats are provided for this purpose and a collar and nut provided. These threads are all right-handed as a milling arbor must be fitted with a taper shank and not driven from the chuck.

This completes the drive mechanism and a lot of useful work can now be done, but to make it really versatile as a horizontal borer, the quartering table, fully graduated, should be made and I will now describe this.

Quartering Table

There are only two simple patterns for the quartering table and no core box. The quartering table pattern has four small lugs, about an inch long and $^{1}/_{4}$" square, glued in the centre of each side and flush with the square face. These are for holding against the faceplate for turning. The baseplate is first machined on the bottom face by holding in the four-jaw chuck. My 6" four-jaw chuck has the body screwed to fit the lathe nose and the casting will just clear the edge of the gap. The turning is best done with a tungsten-tipped tool to as fine a finish as possible and this face is then checked and roughly scraped to the surface plate. The holes for the tee-bolts are next marked out but drilled $^{13}/_{64}$" only. Then, placing the casting on the faceplate, machined side down, a number of these holes are selected that can be used for bolting down and are tapped $^{1}/_{4}$" Whitworth, and Allen cap-screws with plates under the heads used to bolt the casting to the faceplate. The well rounded corners of the casting will nicely clear the gap. Note that the centre of the casting is $^{1}/_{8}$" off centre.

The top face is now machined to exactly 0.005" leaving say $^{1}/_{2}$ thou for scraping both sides. The hole is bored $1^{7}/_{8}$" and the face of the part where the indexing tongue fits turned to just under $^{5}/_{16}$" above the face of the circular part, to clear the four corners of the quartering table. Make sure the unturned rear part of the base casting is below the turned face, or if not, carry the machined face to a greater diameter than $6^{1}/_{8}$". While still on the faceplate and with the spindle locked with my dividing attachment, I milled the 60 deg. groove for the indexing tongue using the same cutter as for the main slides. This was carried down slightly deeper than the already machined surface so that the $^{5}/_{16}$" thick tongue piece would clear the table. I also faced off each side of the raised parts of the casting between which the tongue fits so that I had a true surface on which to screw the fulcrum bracket and a true edge for drilling the gib screws. When milling this 60 deg. groove, allow $^{1}/_{8}$" for the gib piece, so that the tongue is approximately in the centre. Complete the drilling and counterboring for the $^{1}/_{4}$" Allen cap screws as drawing if your lathe is a Super 7, but if not to fit your lathe. Note that the drawing shown allows of three positions on the cross-slide to take full advantage of the 6" of travel. The position shown is the extreme 'out' position. Now hand-scrape to surface

plate both sides of the base, checking that 0.005" is maintained at all points.

The table is first secured to the surface plate by means of four clamps on the small lugs mentioned, packing with card first at these four points and with more if necessary to prevent the casting rocking. Make sure it is central to the four sides. It is now faced on the working face and the spigot turned to a good working fit in the base hole – any play here will detract from the indexing accuracy. The outer diameter of the shoulder is turned to $6^{1}/_{8}$" dia. and the underface of the four corners faced so that the shoulder is $^{5}/_{16}$" high. I made a tee-slot tool from a 4" length of $^{5}/_{8}$" key steel and I have made a special drawing of this as it is a little difficult to explain. The groove for the circular tee-slot is first made $^{3}/_{8}$" wide on a centre diameter of $4^{3}/_{4}$" and $^{7}/_{16}$" deep. Then the left- and right-hand tee-slot tool is used to bring the total width of the tee-slot at the bottom to $^{5}/_{8}$" wide, the grooves being $^{3}/_{16}$" thick, there being a height of $^{1}/_{4}$" left.

I made the 360 indexing divisions with a pointed tool on its side, using the indexing fixture mentioned and a saddle stop for the three lengths of line required, namely, the individual degrees at $^{1}/_{8}$" long, every fifth at $^{3}/_{16}$" long and every tenth at the full length. The two zero and the two 90 deg. marks were made by bringing the casting square to start and the tool 'dead on' centre line. To machine the flat square face of the table, I used the four clamping $^{1}/_{2}$" thick plates of the main slides, interposed between the faceplate and the machined face, using four good tee-bolts. Before these could be used however, I had to cut through the casting a hole, shaped as the drawing, the position of which was determined by drilling through from the circular tee-slot. A number of small holes were drilled through from the flat face and the iron cut away with a small chisel.

The table casting can now be mounted on the faceplate for turning the square face. This should be turned to a good finish to $^{1}/_{2}$ thou above 1" to allow for scraping. The circular side will have to be scraped to the

The quartering table complete.

already true face of the base and when true, the base should be re-scraped for about an inch or so each side of the two holding down bolts, so that a little tension is placed around these holes, making sure that the surfaces left are tightly in contact. The square face is also scraped true to the surface plate and to the micrometer at 1". I realise with horror that I have forgotten to mention the four indexing holes in the circular part of the table! These holes must be drilled with a drilling jig under the toolpost on the exact centre height and with a ¹/₄" drill. The drilling is done with an electric drill on the four zero marks in the centre of the table as close as possible to the square table at the same setting and before disturbing, immediately after the indexing is done, using the indexing fixture.

Now set up the whole fixture, bringing the unmachined edge of the table against the faceplate to bring it square. First however, you will have to make the eight tee-slot tongues and make the eight Allen ¹/₄" cap-screws to the proper length. Make sure that the screws do not bottom in the tee-slots as this may quite possibly pull off the tee-slot edges. The screws must also be just clear of the top of the base. I used ⁹/₁₆" key steel for the tongue pieces, milling a full length and then cutting off at ¹³/₁₆" long. The two ³/₈" BSF teebolts for the circular tee-slot are made from 1¹/₈" round m.s. with the heads to conform to the tee-slot. Having got the fixture set up with the table square, it is necessary to mark out the position of the ¹/₄" spigot on the end of the indexing tongue. Do this as carefully as possible and turn to a slight taper and a good fit in the four indexing holes in the table.

Indexing Gear

The rest of the indexing gear should now be made, before completing the table. One end of the lever is riveted into the fulcrum block and the lever slides in the other block on the indexing tongue, both blocks being free to turn in their places. A spring, between the stop in the base and the lever, is provided. These two stop blocks I find I have forgotten to mention: they are inset $^1/_8$" into the edge of the base as shown and secured with a 2BA cap-screw. They ensure that the whole fixture is brought true to the cross-slide edge.

With the indexing pin in place in one of the four holes, the edge of the table can now be milled. This is done with a facing or fly-cutter and the four edges are milled, preferably to the same saddle setting. I milled the four tee-slots, using a $^3/_8$" end mill and followed with the $^9/_{16}$" dia. tee-slot cutter used for the tables of the main slides, mounting the table on the faceplate again. I had to move the milling fixture along the crossslide for each slot, as the table is rather longer than the cross-slide travel. I was able to mill two of these slots and then index 180 deg. for the other two, getting them parallel with the machined edge of the table by setting the table level with a dial gauge held in the toolpost.

It occurred to me much later that I could have milled these tee-slots in place on the cross-slide on its own base, which would have made for complete accuracy and I could have carried the milling right through without resetting, by mounting the elevating heads on the lathe, removing the headstock and tailstock and mounting a heavy steel slot bar on the two tee-slotted tables over the saddle. Then, by mounting my milling fixture in a vertical position on the middle of this flat bar, I could have milled the tee-slots to any length. I could also have made the quartering table bigger (up to $7^1/_4$" square) and still have turned it in the normal headstock in the gap and have milled these tee-slots. This rather proves to me that my head is useful only for holding on my collar

and also demonstrates the versatility of this fixture.

A zero mark is made on the right-hand edge of the base for indexing other than the four quartering positions.

Setting Up

To set the two elevating heads up in place on the lathe with the chain drive mechanism in place, a position for the chain support bracket where the chains are in line is maintained on the chain support bracket by fitting a small cap screw underneath to butt against the edge of the lathe gap. The headstock is secured by drilling the rear machined face of the Myford bed for the two $^5/_{16}$" Allen cap screws. These screws come lower than those for the taper turning attachment and do not interfere with them. With an electric drill, first spot with a $^3/_8$" drill and then drill $^{17}/_{64}$" and tap $^5/_{16}$" Whit. If you have a favourite milling arbor, the tailstock position can be drilled and tapped in the same way. Then mount a fairly long boring bar in the two heads of at least 1" dia. and turn two diameters to exactly the same size, one on each end of the bar to about 1" long in each case and in such a position that a feeler-gauge can be interposed between these two turned diameters and the cross-slide. Then, by getting the same reading on the headstock and tailstock and knowing the distance to the centre ofthe bar from the feeler gauge, a zero mark can be set on the index collar of each slide and the screws tightened permanently.

Also, the zero mark can be put on each teeslotted table opposite the 2" mark, if setting is done over the cross-slide, and opposite the $^1/_2$" mark, if setting is done above the quartering table. It is much easier if the bar is turned to say 0.020" under 1" and a feeler gauge of 0.010" used: thus the zero mark is made $^1/_2$" above the 2" mark on the slide and opposite the 1" mark for the quartering table. One point to remember when setting the heads to any mark, is to go below the mark and then come up to it before tightening the gib locking screws.

This takes up any backlash in the screws and setting for, say, $5/8$" above the table or cross-slide is then simple. Set first to the 0.600 mark on the side of the slide, with the index collar on the zero mark, and then turn to the 0.025 position on the index collar and lock, and you have it.

My lathe is of the long bed variety and my top support bar is 30" long but this can be made any length to suit. Setting the chains is simple: when the heads are in place at the desired height, hold the two chains taut and tighten up first the bracket support bar, then the slotted bar at the sprocket spindle nut and finally the headstock screw.

This fixture is rather in the nature of a major undertaking, but when completed you will have the satisfaction of knowing that no project within reason is beyond the capacity of your lathe. Using this fixture and my milling attachment and hob-type cutters with the indexing attachment, I recently made, completely on the lathe, a four speed all-geared motorised feed box, complete with reversing switch, for a full size plano-milling machine, and many other projects seemingly impossible for so small a lathe.

CHAPTER 8

A Quick-Change Toolholder for the Lathe Tailstock

Chuck work in the lathe almost invariably means drilling a number of different size holes, tapping, screwing, etc., and if a number of similar parts are required, changing drills in the chuck, removing chuck and fitting dies can become very tedious, besides wasting a lot of valuable time.

I determined at my first opportunity to design a revolving turret for the tailstock. On making a few rough sketches it soon became apparent that with, say, a 1/2" drill chuck and drill in place, difficulty would be experienced in swinging this and other tools around to clear the toolholders on the topslide and to clear the rear parting off toolholder. Even withdrawing the tailstock while swinging the turret around did not solve the problem. Very long tools in the turret might possibly allow the operating tool to approach the work, but this would mean a great deal of overhang, for which the tailstock is not designed.

Accordingly I have designed this Quick Change Toolholder for the tailstock which takes up no more room than a single tool and works with a minimum of overhang.

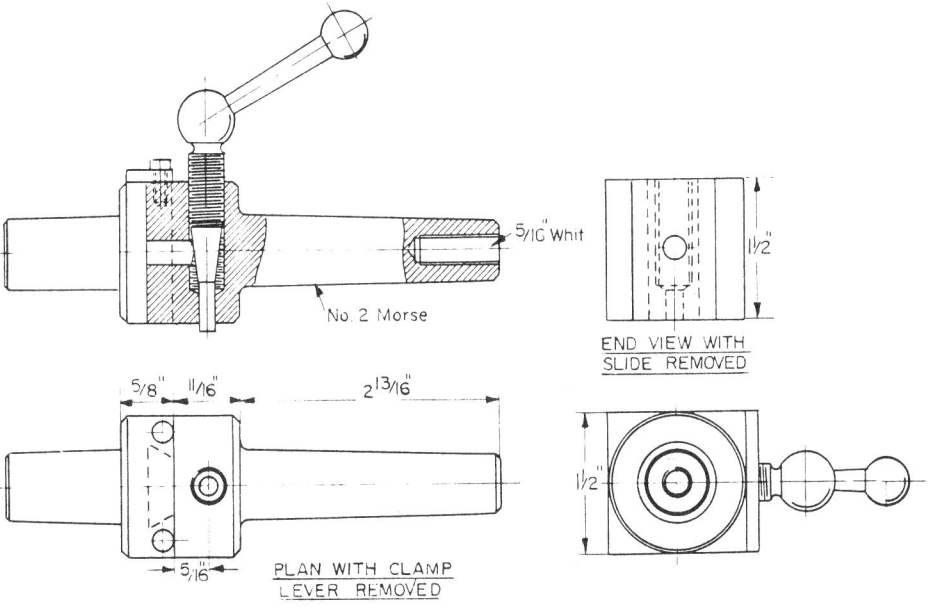

5/16 Whit

No. 2 Morse

1/2"

END VIEW WITH
SLIDE REMOVED

5/8" 11/16" 2 13/16"

1/2"

5/16

PLAN WITH CLAMP
LEVER REMOVED

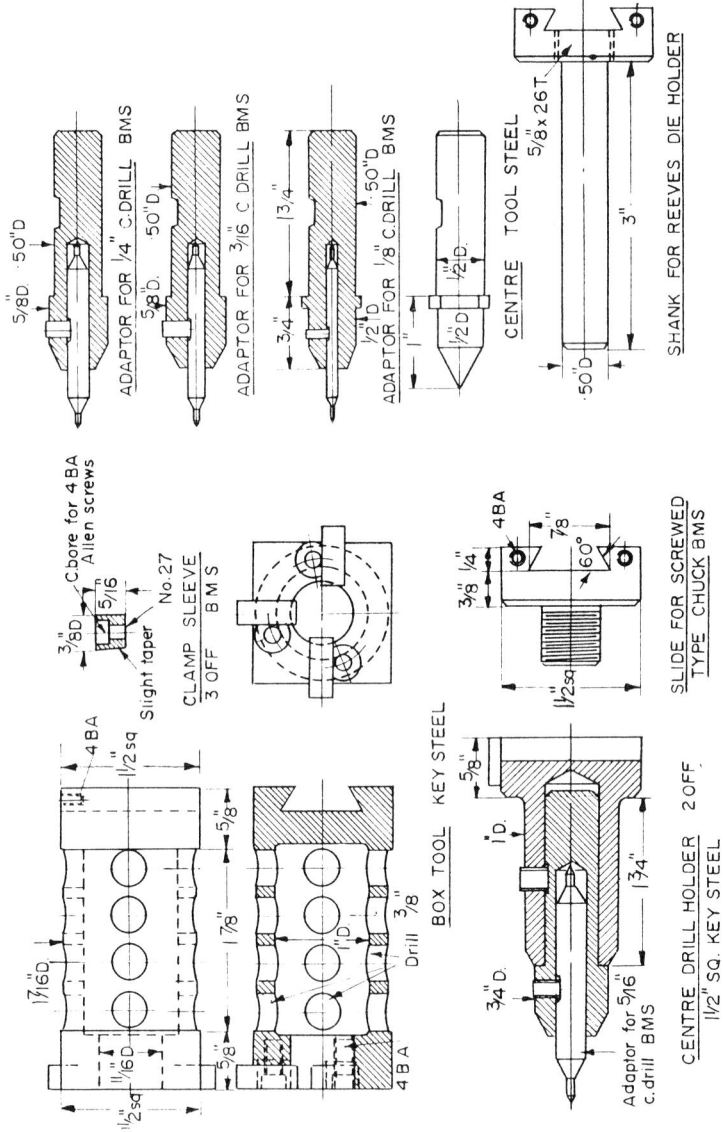

Operating the fixture is almost as quick as a turret and there is no limit to the number of tools which can be fitted. The tailstock does not need to be moved along the bed and thus the readings for depths of drilling, etc., will apply to subsequent parts being made. I have shown nine toolholders or 'slides' but there is no limit and if one wants to go into the question of box tools for repetition work the field is enormous.

First there is a centre drill holder with four different size adapters for $1/8$", $3/16$", $1/4$" and $5/16$" dia. centre drills. One chooses the nearest size of centre drill that is suitable for the particular job in hand and fits this complete with its own adapter in the centre drill holder slide. Next there is a slide for a $1/4$" drill chuck with its own chuck permanently fitted and also a slide for a $5/16$" chuck, one for a $3/8$" chuck and one also for a $1/2$" chuck, each with its own chuck fitted. Then there are two die holders, one for the Myford die holder and one for the Reeves type as it often happens that two threads must be cut on a part being made. I have also included a slide with a lathe centre fitted as it is often necessary to support work with the centre during part of the operation. This centre is parallel and is easily removable so that a different type of centre, such as a half-centre can be fitted. Lastly, a simple type of box tool which will turn long, slender work in the chuck without centre support and to accurate diameter. This could be elaborated with limits only on the requirements of the work and the ingenuity of the designer.

There is only one shank and in operation this is knocked into the tailstock and, having the slides on a bench close to the left-hand and in order of operation, each slide is pushed down into contact against the stop plate fitted to each slide and tightened up with the clamp lever. If each slide is in its proper order on the bench and each position is marked for operating depth, very little time is lost in changing the tools.

This fixture is made mainly of $1^1/2$" key steel or $1^1/2$" square b.m.s. and about 21" will be required for the nine heads; if a couple of feet is purchased, an additional end can be milled for a subsequent slide. A dovetail cutter, $7/8$" largest diameter and 60 deg. angle is required. If this cannot be purchased, it is better to make it of H.S. tool steel, the shank should be $1/2$" dia. and 10 teeth will be right. The outer large face should be relieved in the centre and shallow teeth cut across. Also required is a $5/8$" end mill and a facing cutter about 2" dia. These three cutters are gripped in the three-jaw chuck in turn; for milling the end of each slide, the pieces of $1^1/2$" square M.S. are held in turn on the cross-slide with the end facing the chuck for milling.

Two pieces $2^1/2$" long are cut off, one for the centre drill holder and one for the centre holder. This will leave sufficient for facing each end. One piece $2^1/2$" long will take care of the chuck holder for the $1/2$" and the $1/4$" chucks, as these are best cut apart after milling, and one $2^3/8$" piece will take care of the other $1/4$" (or $5/16$") chuck and the $3/8$" chuck. The two die holder slides are only $5/8$" finished length, so these are milled each on the end of the bar remaining and cut off and, finally, the box tool piece is milled and cut off at $3^3/16$" long to finish to $3^1/8$".

These nine slides must be milled with the dovetailed slide right in the centre of the piece and dead square across, so two packing pieces lengthwise with the cross-slide are made up and a piece to act as a stop is bolted down square so that the pieces can be butted against this stop piece to bring each piece square. Get the stop piece square to the faceplate in the first place with a good square. Then, with the saddle locked in place, face with the facing cutter across, feeding from the rear to the front edge so that pressure is downwards. Now with the $5/8$" end mill, mill to a depth of 0.255" right in the centre, right across, using the lead screw graduations to secure this depth. Then change the cutter to the dovetail one and, touching against the face to get a reading, mill to a depth of 0.250" exactly. This will leave a 0.005" relief in the centre of the dovetail slide. Mill all the

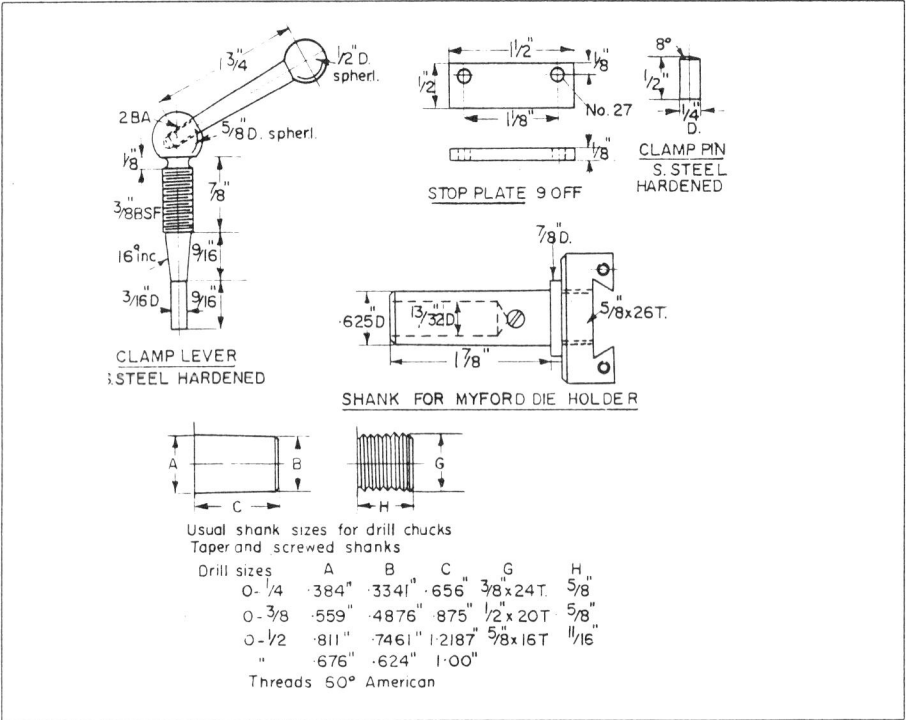

CLAMP LEVER
S. STEEL HARDENED

STOP PLATE 9 OFF

CLAMP PIN
S. STEEL
HARDENED

SHANK FOR MYFORD DIE HOLDER

Usual shank sizes for drill chucks
Taper and screwed shanks

Drill sizes	A	B	C	G	H
0 - 1/4	·384"	·3341"	·656"	3/8 x 24 T.	5/8"
0 - 3/8	·559"	·4876"	·875"	1/2 x 20 T.	5/8"
0 - 1/2	·811"	·7461"	1·2187"	5/8 x 16 T.	11/16"
"	·676"	·624"	1·00"		

Threads 60° American

pieces and perhaps the end of the remaining piece in the same manner. A speed of 200-290 r.p.m. will be about right and use plenty of cutting oil with the cross-slide closely adjusted. This should result in all of the slides being the same size and to a good finish.

A piece of the same 1½" square steel will be required for the main shank, cut off 3⅞" long. This is milled with the same three cutters but this time it must be set up on the vertical slide. The slide must be set truly square with the faceplate and the steel bolted securely square also.

A good deal of skill is required to make a properly fitting dovetail to fit all the slides and to finish upright in the centre so that the sides of the pieces are flush with each other. Mill 5/16" wide off each side to a depth of, say, 0.245" after facing off the end, and

then using the dovetail cutter, mill carefully in top and bottom to a depth of 0.250" from the end. A reading both top and bottom on the vertical slide, taken by letting the cutter touch top and bottom face before commencing milling, will help to centralise this milling. Get the fit to a good close slide fit on the slides.

The next operation is to drill and tap for the clamp screw before any turning is done on the shank. Bolt down on the drilling machine table over one of the slots in the table and drill 3/16" right through, exactly in the centre, and 5/16" from the lower milled face. Then drill 21/64" to a depth of 1¼" only. The drilling is parallel to the milling as shown on the drawing. This is now tapped 3/8" BSF.

Now set up in the four-jaw chuck using the dial gauge and, traversing the gauge along each side in turn, bring the readings to the same for each side and for the full length. The indexing head should be used here to bring each face square with the dial. The milled face is into the chuck. Centre the unmachined end rather deeply and face up, using the half centre.

Then, using the normal centre, proceed to turn the length No. 2 Morse taper for a length of 2¹³/₁₆". Get a proper full length fit and to a good finish.

The end is next drilled ¹⁷/₆₄" and tapped ⁵/₁₆" Whit. for a depth of 1". The end should be counterbored ⁵/₁₆", slightly, and finished inside to 30 deg. to finish off professionally. The outer edge is finished off with a round-ended lathe tool.

All the slides are machined in place on the main shank, with the shank drawn into the headstock of the lathe, but before this can be done the clamp screw, clamp pin and nine stop plates must be made. The clamp screw is made from silver steel and is reduced to ³/₁₆" for ⁹/₁₆" long to fit the drilled hole in the main shank and then tapered 16 deg. inclusive and the balance to a total length of 2" is screw-cut ³/₈" BSF. Cutting off at 2⁵/₈", the end is spherically turned to a ⁵/₈" sphere, holding the piece in a screwed sleeve in the chuck. Drilling and tapping 2BA at 30 deg. through the centre for the clamp lever, the clamp screw can now be hardened in oil and tempered to a light blue on the thread and to a straw colour on the taper and the end. The dome end is polished and the lever made with a ¹/₂" spherical end in b.m.s. to bring to 1³/₄" centres as the drawing, slightly tapering the lever.

The nine stop plates are made from ¹/₂" × ¹/₈" b.m.s. and are 1¹/₂" long with two No. 27 holes at 1¹/₈" centres ¹/₈" from one edge for the two 4BA Allen cap screws. These should be jig drilled using a spare plate. The nine slides can now be drilled and tapped 4BA for these stop plates and these can be fitted to the slides.

Before turning the slides there is still a number of things to do. You will notice that

for the two die holders, the shanks are screwed into the slides to save a lot of turning and waste of steel. The two slides are held truly in the four-jaw chuck and bored and screw-cut and tapped ⁵/₈" × 26 t.p.i. and the two shanks are made from ⁷/₈" b.m.s. and screw-cut to fit. These are then fluxed on the threads and silver soldered into position. The four centre drill adapters are turned 0.500" for 1³/₄" long to fit a ¹/₂" reamed hole and the centre of ³/₄" tool steel is turned to the same dimensions. The four centre drill adapters are parted off at 2¹/₂" long and the centre is parted off 2³/₄" long. A small flat is filed on the ¹/₂" dimension to take a ¹/₄" Allen grub-screw at ⁷/₈" from the shoulder.

It will be necessary to purchase the four chucks before turning the chuck holder slides as the tapers or threads used by the makers vary greatly for the same size of chuck. I purchased a ¹/₄", a ⁵/₁₆", a ³/₈" and a ¹/₂" chuck, all with tapered holes as I prefer this type for accuracy. I have listed some of the commonly used tapers and threads but do not rely on these; it would be better to turn a short piece of round b.m.s. in the chuck to get a proper setting for the topslide but the taper for each size of chuck varies greatly.

Now the main shank can be drawn into the headstock spindle of the lathe with a draw bolt and the end centred and drilled D and reamed ¹/₄" for the clamp pin. This pin, by the way, is hardened after turning the end to 98 deg. to conform with the taper on the clamp screw, the other end being flat with corners very slightly rounded. Then with the clamp screw and clamp pin in place, slide the centre drill holder slide into place firmly against the stop plate and clamp up with the clamp lever. Unscrew the lever from the clamp screw as it may be in the way during turning, and centre the end of the slide and turn to 1" dia. for a length of 1³/₄" leaving ⁵/₈" thickness on the 1¹/₂" square part. Drill ³¹/₆₄" for 1⁷/₈" depth and bore to 0.495" and ream ¹/₂".

With reference to reaming in the lathe, I

A QUICK-CHANGE TOOLHOLDER FOR THE LATHE TAILSTOCK

have noticed that hand reamers are often used in articles in *Model Engineer*, the squared end being held with a spanner against the tailstock centre. This is very bad practice and hand reamers are one tool that I would not waste money on. Machine reamers, taper shank spiral flute, have only a very short chamfer on the end and the end is as much as 0.0005" larger at the tip than it is at the other end of the flutes. The flutes are also irregularly spaced and fluting is done left-handed. With a very slow speed and a very fast feed, straight in as fast as possible, pause and straight out again, the result is first class, smooth, true to size and parallel, the bottom of the hole being the same size as the outer edge.

The outer edge of the centre drill holder is turned to 30 deg. setting on the topslide leaving a $1/8$" wide shoulder on the end and after fitting the $1/4$" Allen grub-screw at $7/8$" from the end, the first centre drill adapter can be slipped into place and tightened down. This is then faced, centred and drilled $5/16$" for the largest centre drill, turned outside to $3/4$" and chamfered to 30 deg. to leave a small flat on the end.

The adapter for the $1/4$" and the $3/16$" centre drills are done in the same way, except that these two are reduced to $5/8$" dia. on the end and the adapter for the $1/8$" centre drill is reduced to $1/2$" but leaving a $1/8$" wide shoulder $5/8$" dia. A small flat is ground in the centre of each centre drill and an Allen grub-screw fitted to each adapter.

The slide for the centre is machined in exactly the same way and to the same dimensions. The centre itself is fitted and the end turned down to $1/2$" leaving the same shoulder as the smallest adapter, and the end is turned to the 30 deg. topslide setting. The shoulder has two flats to take a spanner so that the centre can be removed. It is then hardened in oil and tempered slightly from the shank to the point.

The reason for making these centre drill holders long is so that the tailstock movement will be more or less the same for different drills, allowing the tailstock to

remain clamped in one position.

The two die holders are mounted on the main shank in the same way, centred and turned and finished to the same dimensions, etc., as your ordinary die holder shanks, leaving a $1/8$" wide shoulder against the slide.

The four chuck slides can now be machined to fit your chucks, leaving $1/16$" between the face of the slide and the back face of the chuck, if of the tapered hole type, and to butt against the face of the slide, if of the screwed type. The last $1/8$" of the thread could be undercut to facilitate threading right up to the shoulder in the screwed type. In all of these slides, with the exception of the box tool, the face of the slide is turned to $1 1/2$" dia. to about $1/8$" deep with a round-nose tool for appearance.

This leaves only the box tool. This is set up, faced and centred and the outside diameter is turned in the centre part to about $1 7/16$" leaving $5/8$" width at each end the full $1 1/2$" square. This is then drilled $5/8$" to $2 1/2$" depth and bored $11/16$" to clear a $5/8$" job. It is then bored out inside to within $5/8$" of each end, to $1 1/16$" to clear turning swarf. Four $3/8$" holes at $1/2$" centres are drilled into the centre bore from each of the four sides, so that swarf can be easily removed and the work seen and, as can be seen on the drawing, there is one $1/4$" square tool and two $1/4$" square stops all made from $1/4$" tool bit pieces. The top of the tool is on the horizontal centre line while the two stops are central to the centre line. Before milling these $1/4$" × $1/4$" slots, it is necessary to drill the three flat bottomed $3/8$" holes, at $3/8$" depth and $5/32$" centres, from where the edge of the tool or stop will be.

These holes are drilled right through and tapped for 4BA headed Allen screws. The clamp pieces are $3/8$" dia. × $5/16$" long, drilled No. 27 and drilled $7/32$" and flat bottomed to $5/32$" depth to accommodate the head of the 4BA Allen cap screws. A small tapered flat is filed on one side of each clamp piece (or collar as it could be called) so that on screwing down into the hole, it tightens against the tool or stop.

I have used this method of clamping tools when making facing cutters for milling and it has proved very satisfactory. The $1/4$" slots for the tool and the stops can be end milled on the cross-slide or radial milled on the vertical slide, the latter method being perhaps the better for getting a close fit. In using the tool, the cutting edge is set very slightly ahead of the stops so that true running work will result, the stops can be slightly chamfered to achieve this. Other small box tools with several tools and stops in sequence along the length could be made to turn several diameters at once if desired. There is no limit to the variety of tools that could be made.

In use, this fixture has proved well worthwhile and is a great time saver, even for a two-off job time can be saved. So convenient is it that I leave the shank in the tailstock at all times ready for use, removing it only for heavy between-centre turning, boring and milling. The draw bolt does not need to be used in the tailstock, knocking in being quite sufficient.

CHAPTER 9

A Toolpost Grinder

There have been many requests from enthusiasts for particulars of my toolpost grinder, photographs of which have appeared occasionally in various magazines.

The chief advantage of my particular design is that it is made specifically for the Myford lathe, the grinding wheel can be exchanged end for end with the pulley, making it left and right handed so that a Morse taper for instance can be ground on the end of a shaft with the taper up to the tailstock and, when reversed, the lathe centres can be ground, milling cutters sharpened, etc., while when the toolpost grinder is mounted on the vertical slide with a cup wheel, such things as facing cutters and angular cutters can be ground.

This fixture uses the same motor, motor bracket, pulley and belt as my Milling Attachment described in Chapter 1, and is a series wound, $1/8$ h.p., 4,500 r.p.m., 1.0 amp. continuous rated motor, made specifically, I think, for driving fans. It has an end mounting flange which makes mounting easy. The belt is a sewing machine belt $1/4$" wide of V-section.

Main casting

The casting is of iron and the pattern is made mainly of $1/2$" plywood, with a piece of $1/8$" ply glued on the bottom for machining, leaving $2^5/8$" of this undersurface relieved $1/16$" as shown on the side elevation. A core box is required, but this is made from two simple pieces of wood $2^1/8$" thick shaped out where they fit together for the core. The prints for this slotted hole are $1/2$" high on each side of the main pattern. The core prints for the spindle hole are $5/8$" dia. and should be $3/4$"

The author's toolpost grinder mounted on the vertical slide for sharpening a rack cutter.

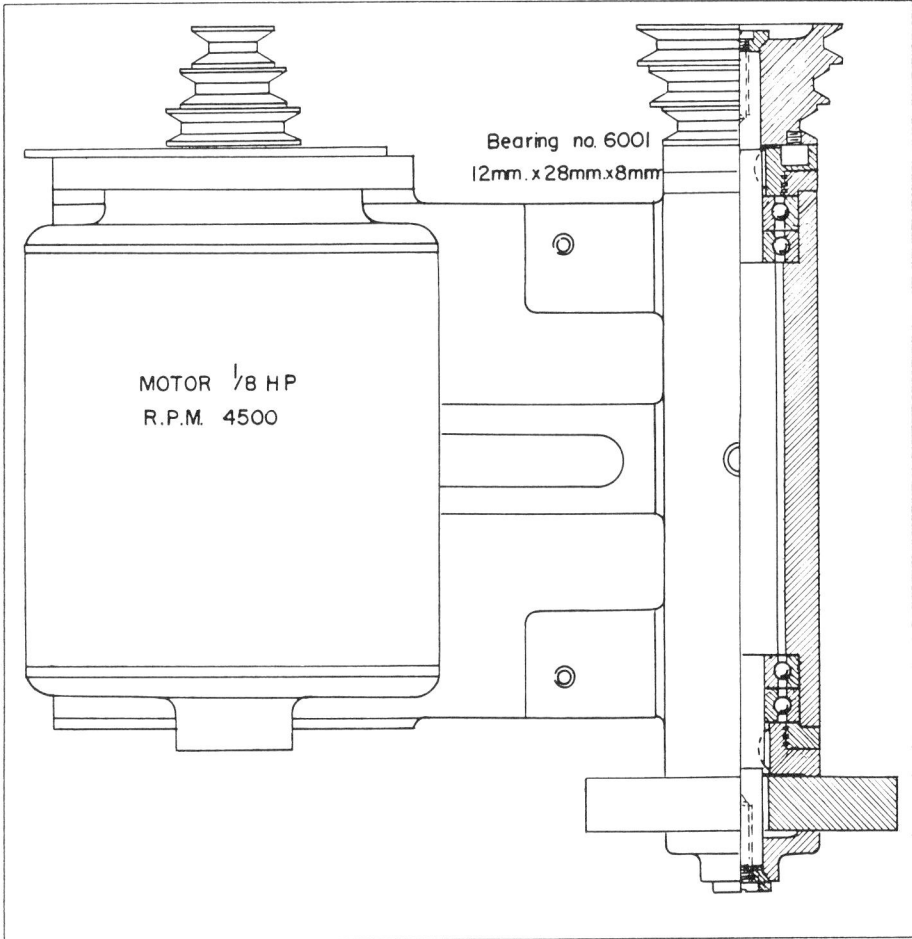

Bearing no. 6001
12mm. x 28mm.x8mm

MOTOR 1/8 H P
R.P.M. 4500

long. Leave $1/16$" on all machine surfaces. I glued a piece of $1/8$" ply, $1/2$" wide right along on top of the spindle housing. This was milled true first to give me a flat surface for bolting against the faceplate for turning the back, the other end of the casting being packed out to bring parallel. Four $3/8$" tapped holes were drilled as shown on the No. 2 plan view so that I could bolt on to the faceplate. This surface was turned flat and hand scraped to the surface plate.

I next re-mounted the casting back on the faceplate with the machined face against it and milled the top face of the cored slot, the top edge of the motor mounting edges and the wheel guard pads, with an end mill mounted in my milling attachment, the spindle being locked on the lathe.

Then, packing up on the cross-slide and with a large diameter facing cutter, the two sides of the castings were machined to 5" wide, making sure that these faces were faced parallel and square with the spindle housing.

Now mount the casting on the topslide, bringing it dead square by butting against the faceplate. First, however, I made the special packing washer with its spherical concave face to fit the Myford toolpost washer and used this for bolting down.

A $1/2$" boring bar was used to rough out the bore for the spindle and this was followed as soon as possible by a $3/4$" boring bar and the bore was taken out to $7/8$". This was followed by boring each end to a very light push fit for the bearings leaving, by very careful measurement, $3^5/8$" of the original $7/8$" dead in the centre. The bearings are four off No. 6001 and are 12-28-8mm or 1.1024"-0.4724"-0.315", they are of the deep row rigid type, butted two side by side, and are better than the self-aligning type as they will stand a certain amount of side thrust. This boring should leave 0.057" after the bearings are in, or in other words, should be carried to a depth of 0.687" each end, provided that the body is exactly 5" wide. As stated, a very light push fit or a sliding fit is all that should be provided. The shoulders should be square in the bottoms of the holes.

The spindle

The spindle is a piece of $3/4$" b.m.s. faced off each end to $7^1/2$" long and centred. It is turned to approximately $11/16$" and a little more than a stiff push fit for the bearings. Tightly fitted bearings can distort and result in premature wear.

The centre part should be exactly the same length between the shoulders as the centre part of the bore in the casting. This is most important, but to get this right is not difficult. Make a collar with a true face to slide on the shaft and to slide in the bearing housings and butt this collar against the shoulders on shaft and housing, then with a depth gauge, measure the other end of housing depth and bring the shoulder on the shaft to the same depth. This will result in a

very satisfactory bearing design, with no side thrust on bearings and no end play in the shaft, it will be very free running and will last indefinitely. The shoulders on the shaft should be square. The ends at $7/8$" long are turned $7/16$" to a nice slide fit in a $7/16$" reamed hole. The ends are tapped 2BA for the holding collets. A $1/2$" dia. × $1/8$" wide Woodruffe cutter is required for the keyways and this should be sunk in at $7/8$" from the shoulder, to the centre of the keyway.

The two end covers are made from 2" dia. b.m.s. and the boring should be done off-centre by the same amount the housing is off-centre, which will be approximately $1/8$" and can be measured from the bored housing. The bore has three tiny oil grooves machined in which form an effective oil seal. The depth of the shoulder at 0.057" should be checked with the depth gauge in the housing and there is no need for gaskets with truly faced surfaces. Make a little groove in the face at $3/4$" radius for the four screws.

The end covers are now put in place in the housing and marked all around the square of the housing and they can then be sawn and filed to a nice flush fit with the outside of the housing. A line scribed from corner to corner of the end covers will give the positions of the four screws which should be drilled and counterbored for 4BA Allen cap screws, the heads of which should be just below the surface.

Driving flanges

The driving flanges made from $1^1/2$" b.m.s.; they have the same bore as the bearings and a $1/8$" keyway about $3/32$" deep to clear the Woodruffe keys in the shaft. The $7/8$" shoulder should clear the bore of the end covers by no more than 0.003" to 0.005" and again, three little grooves are turned in this shoulder. The length of the shoulder should bring the face of the driving flange to clear the end covers by no more than 0.005". A $1/4$" hole is sunk $3/16$" deep with a flat end at $17/32$" from the centre in the outer face of the driving flange and this face

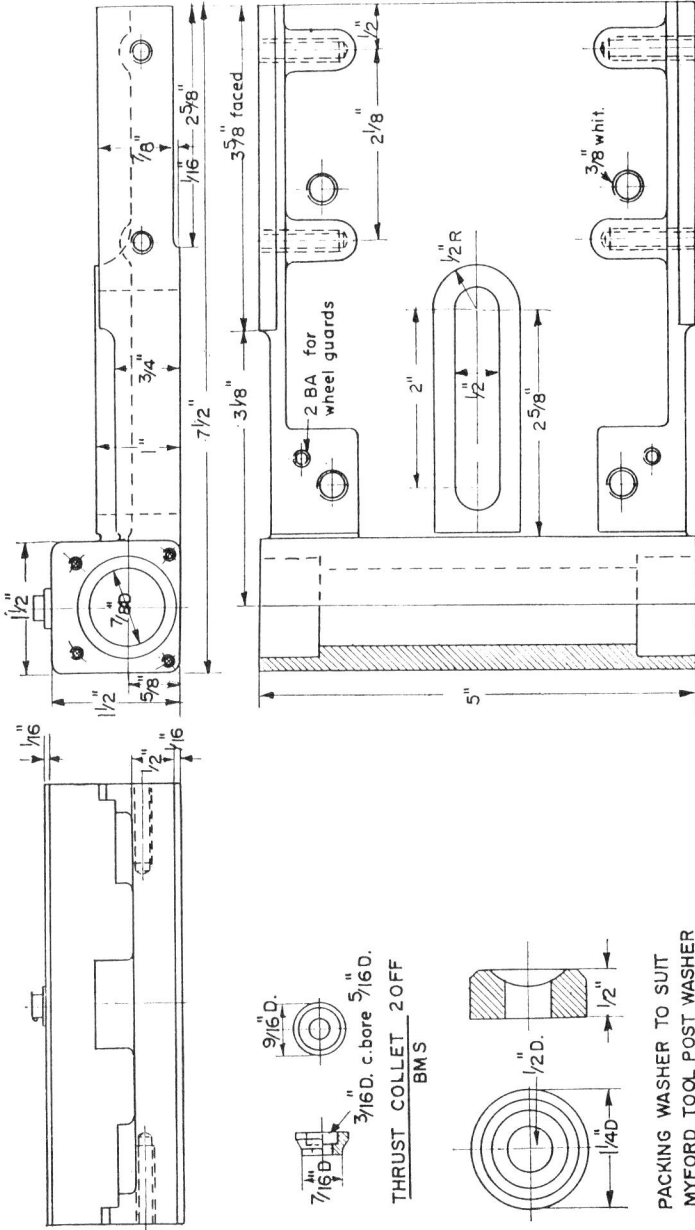

2 BA for wheel guards

3⁷⁄₈ faced

2⁵⁄₈"

⁷⁄₈"

¹⁄₁₆"

³⁄₄"

7¹⁄₂"

3¹⁄₈"

1"

2¹⁄₈"

¹⁄₂"

3⁄₈" whit.

¹⁄₂ R

2"

¹⁄₂"

2⁵⁄₈"

5"

1¹⁄₂"

⁷⁄₈"

¹⁄₂"

5⁄₈"

¹⁄₁₆"

¹⁄₂"

¹⁄₁₆"

9⁄₁₆" D.

3⁄₁₆ D. c.bore 5⁄₁₆ D.

7⁄₁₆ D.

THRUST COLLET 2 OFF
BMS

¹⁄₂"

¹⁄₂ D.

1¹⁄₄ D.

PACKING WASHER TO SUIT
MYFORD TOOL POST WASHER

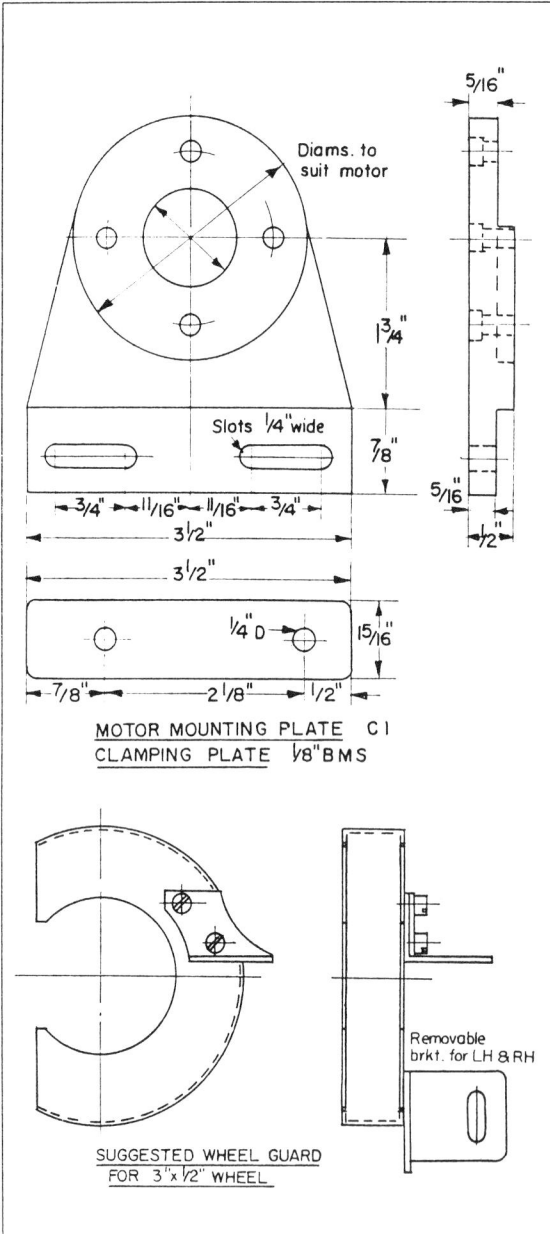

Diams. to suit motor

$1\frac{3}{4}$"

Slots $\frac{1}{4}$" wide

$\frac{7}{8}$"

$\frac{3}{4}$" — $\frac{11}{16}$" — $\frac{11}{16}$" — $\frac{3}{4}$"

$3\frac{1}{2}$"

$3\frac{1}{2}$"

$\frac{1}{4}$" D

$1\frac{5}{16}$"

$\frac{7}{8}$" — $2\frac{1}{8}$" — $\frac{1}{2}$"

$\frac{5}{16}$"

$\frac{5}{16}$"

$\frac{1}{2}$"

MOTOR MOUNTING PLATE C I
CLAMPING PLATE $\frac{1}{8}$" B M S

SUGGESTED WHEEL GUARD
FOR 3"x $\frac{1}{2}$" WHEEL

Removable brkt. for LH & RH

is relieved $\frac{1}{32}$" deep to a dia. of 1".

Thrust collet

The two little thrust collets, which I have so named for want of a better term, are $\frac{9}{16}$" dia. shouldered down to $\frac{7}{16}$" dia. at 30 deg. per side (60 deg. inclusive), and are drilled for 2BA and counterbored to fit the head of the 2BA screws. These make a neat and secure means of holding the pulley and various wheel flanges in position.

The pulley made from 2" b.m.s. is for $\frac{1}{4}$" wide V-belt, outside diameters being 2", $1\frac{3}{4}$" and $1\frac{1}{2}$", the angle being 40 deg. inclusive; each step is a total of $\frac{3}{8}$" wide, the whole pulley therefore is $1\frac{1}{8}$" wide. It is bored and reamed $\frac{7}{16}$" and carefully chamfered on the outside of the bore to 30 deg. setting on the topslide to fit the thrust collets after undercutting this outer face $\frac{1}{8}$" to $1\frac{1}{2}$" dia. In the smaller face, at $\frac{17}{32}$" from the centre, a 4BA Allen cap screw is fitted. This is the screw that transmits the driving force from the pulley to the driving flanges, and through the key to the shaft.

The 3" × $\frac{1}{2}$" × $\frac{1}{2}$" grinding wheel has a $1\frac{1}{2}$" dia. end flange chamfered in the bore the same as the pulley, other dimensions will become obvious. A sleeve $\frac{7}{16}$" × $\frac{1}{2}$" is required or the wheel can be filled with lead and bored $\frac{7}{16}$" while held in the three-jaw chuck. Paper discs should always be used each side of all grinding wheels. I have drawn two other suggested wheel mountings, each of which has a 4BA driving screw. The other dimensions will suggest themselves. A wheel cover is also drawn.

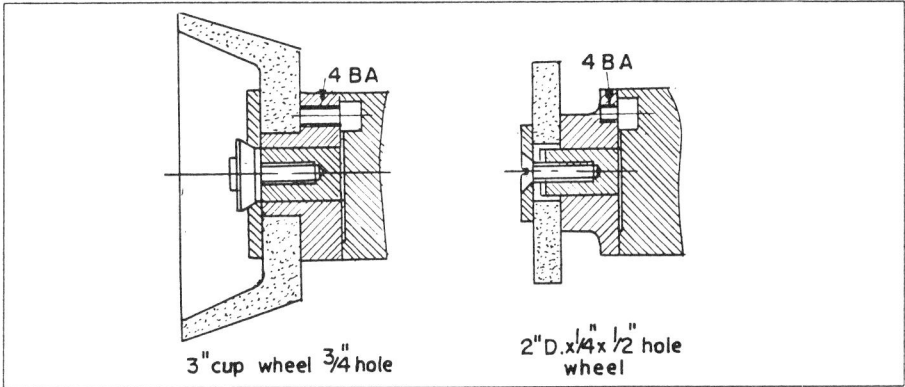

3" cup wheel 3/4" hole

2" D. x 1/4" x 1/2" hole wheel

The motor mounting bracket is also a casting bored and drilled to fit the motor, and a step is milled to ride on the milled edge of the main casting. The holes are slotted as shown to provide adjustment to the belt and a 1/8" b.m.s. plate drilled as shown clamps the whole securely by means of two 1 1/4" × 1/4" Allen capscrews.

For grinding shafts, tapers, etc., an industrial diamond is the best for dressing the wheel, the diamond being held fixed in position in a holder clamped to the tailstock spindle, but for sharpening cutters a piece of carborundum held in the hand is the best for free cutting without burning.

I use the nozzle of a vacuum cleaner held close to the wheel to draw off all dust, particularly while dressing the wheels, and use the coolant for grinding shafts. The bed and all slides, etc., are covered with well oiled rags at all times while grinding.

One other point, do not use grease. A cycle oil cup is fitted to the centre of the housing and light spindle oil is used very sparingly.

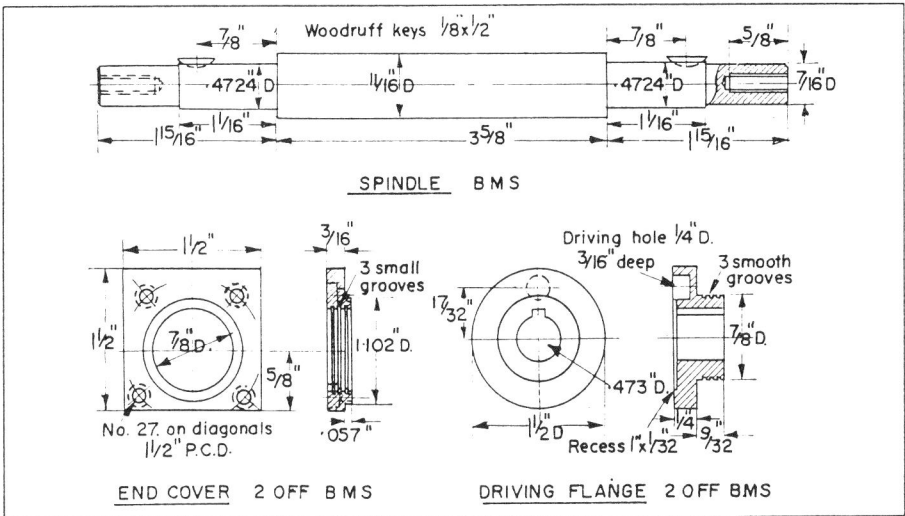

SPINDLE   B M S

END COVER 2 OFF B M S

DRIVING FLANGE 2 OFF BMS

CHAPTER 10

A Tapping Attachment

Tapping in the lathe where the tap is held in the drill chuck in the tailstock and the work is pulled around while held in the headstock chuck is a hazardous procedure in the smaller BA sizes, and is a strenuous and slow procedure in the larger Whitworth and other sizes. While I have not yet broken a tap by these methods, I have been filled with apprehension while tapping down to 12BA. The difficulty of pulling around the chuck while tapping, say, $^1/_2$" Whit. and at the same time preventing the drill chuck from turning in the taper of the tailstock, with the possibility of scoring the taper, has left a lot to be desired as an efficient workshop practice.

I have, therefore, designed two tapping attachments which take care of all sizes and threads from 12BA to $^1/_2$" Whit., for use on the two screwing slides of my quick change attachment. They will fit also the normal screwing shanks or die-holder shanks with No. 2 Morse taper; the advantage of the quick change fixture is that the tailstock does not need to be withdrawn to remove the tapping attachment, merely release and lift straight up.

The smaller of the two tappers has two collers, one $^1/_8$" and the other $^3/_{16}$" and has a knurled shank for tapping the smallest sizes

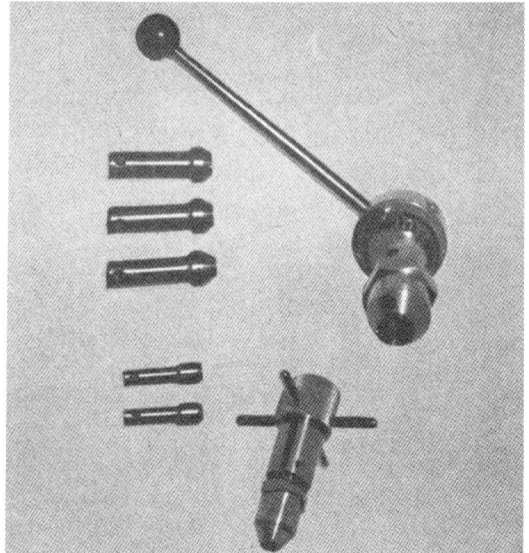

The two tappers with five collets.

and a four-pronged star for tapping the larger BA sizes from 5BA to 2BA and including $5/32$" and $3/16$". The larger tapper has three collets, the $1/4$" size takes care of No. 1 and No. 0 BA and $7/32$" and $1/4$" and the $5/16$" size adjusts to all those taps with shanks above $1/4$" and up to $5/16$" and the largest collet is $3/8$" size which will handle taps with shanks from $5/16$" to $3/8$" which includes $1/2$" taps.

This larger tapper has a single lever which swings to the right to clear the clutch teeth and in the right angled position the clutch teeth are engaged so that the tap can be engaged in the work, the tap advanced and relieved to clear chips, then swung to the right. Further clutch teeth are then engaged so that the tap can be further advanced. The work meanwhile is held stationary in the headstock chuck by engaging the back gear or mandrel lock, and there is no twisting motion on the tailstock as the fixture turns freely on its spigot, thus relieving the tailstock key and keyway of a lot of wear.

All the parts are made of mild steel with the exception of the five collets which are made of oil hardening, non-shrink tool steel and the two phosphor-bronze bushes which fit the shanks of the two die-holders, one being $1/2$" and the other $5/8$".

For the main body of the smaller tapper, a piece of $1 1/4$" free cutting b.m.s. was cut off and faced each end to $3 1/4$" long, centred while held running true in the four-jaw chuck on one end only and drilled $23/64$" for a depth of 3" only, then redrilled to $9/16$" for a depth of 2" and bored to 0.620" and machine reamed to the bottom of the counterbore, namely 2" to the end of the reamer. The balance of the hole, still $23/64$" and 1" long, was bored to run true and reamed $3/8$" which was about the maximum length of the reamer.

A freshly turned stub mandrel $5/8$" dia. to tightly fit the $5/8$" reamed hole was made ready in the three-jaw chuck and the piece tightly twisted on. The end of the piece was now centred and the part turned to the drawing, namely, down to 1" dia. for a length of $2 1/4$", leaving 1" of the full size, which was very lightly skimmed to clean up, and then the end was turned and screw-cut $5/8$" × 26 t.p.i. for a length of $3/4$" without undercutting the end of the thread. The dies were now run over the thread to size it. The centred end was drilled $23/64$" and bored to a depth of $3/4$" only to 0.380". The $1 1/4$" end was knurled with a medium knurl. The four holes for the star spokes were marked out with the dividing attachment before removing from the lathe, by means of a square drilling bush under the toolpost and the electric drill. After removing from the mandrel, it was transferred to the drilling machine and the four holes drilled into the centre, reamed $7/32$" dia. and tapped $1/4$" BSF, as also was the

A TAPPING ATTACHMENT

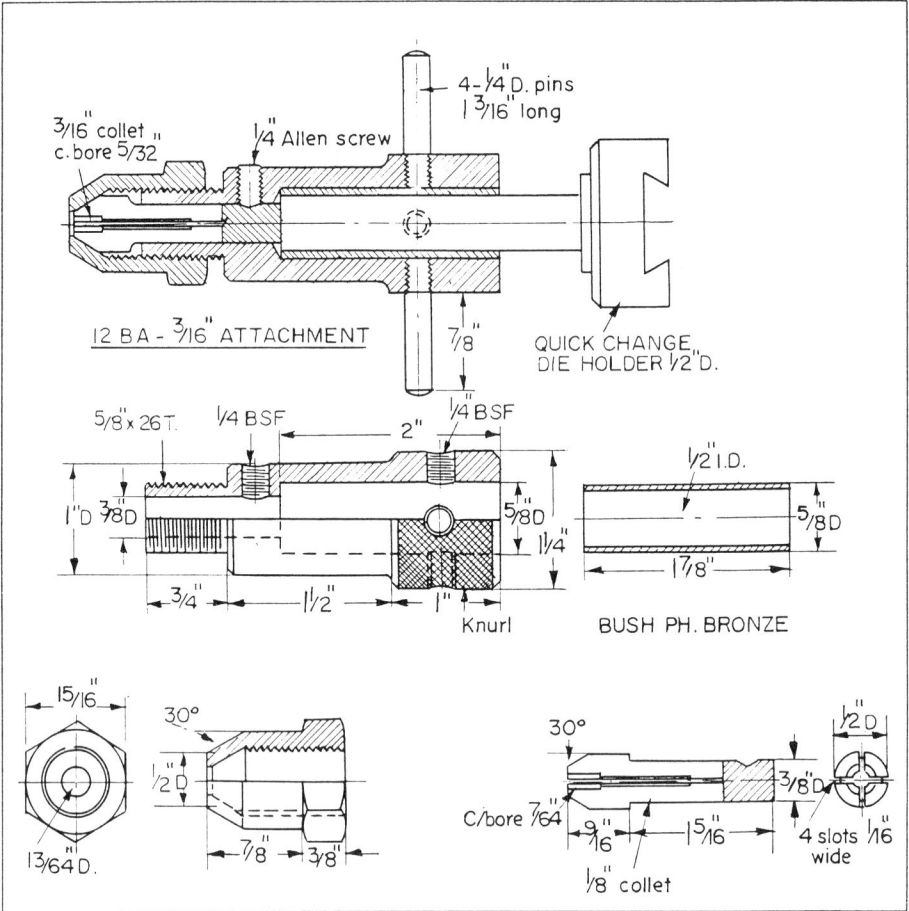

3/16 collet c.bore 5/32

1/4 Allen screw

4 - 1/4" D. pins 1 3/16" long

QUICK CHANGE DIE HOLDER 1/2"D.

12 BA - 3/16" ATTACHMENT

7/8"

5/8"x 26 T.

1/4 BSF

2"

1/4 BSF

1/2" I.D.

5/8D

1"D 3/8D

5/8D

1 1/4"

1 7/8"

3/4"

1 1/2"

1"

Knurl

BUSH PH. BRONZE

15/16"

30°

1/2 D

30°

1/2 D

13/64 D.

7/8" 3/8"

C/bore 7/64

9/16"

5/16"

3/8"D

4 slots 1/16 wide

1/8 collet

grub-screw hole a 1/4" from the shoulder.

The phosphor-bronze bush is 1 7/8" long and is turned outside to 0.0015" larger than the reamed hole and bored to 0.501" as it will close down this much on pressing in. After removing drilling burrs from the 5/8" reamed hole and fitting the four star spokes, which should remain clear of the reamed hole after tightly screwing in, the bush can be pressed home, which will leave a space for oil at the end of the bush. The bush can now be reamed by holding the piece in the

hand with the reamer in the lathe or drilling machine.

I had a piece of 15/16" A/F hexagon steel left over from a previous job, which I used for the tightening nut, but this could be made from round b.m.s. with a flange and tommy bar hole or holes, although I prefer the hexagon steel and spanner. This was drilled right through 9/32" and bored to a depth of 7/8" to 37/64" and, with the topslide set at 30 deg., the end of the hole was bored until the angle met the full diameter of

96

the hole. The bore was then screw-cut 26 t.p.i. and the plug tap run in to size the hole. Then parting off at $1^5/_{16}$" long, a screwed stub mandrel was made and the nut screwed on and turned outside to just remove the hexagon leaving $^3/_8$" and the end chamfered to 30 deg. and faced off so that the end dia. was $^1/_2$" and the length was $1^1/_4$".

With the exception of the collets which we will leave until the last and make them all together, this completes the smaller tapper.

The large tapper

The large tapper main body has to be case hardened on the clutch teeth and also on the 1" dia. journal for the fulcrum sleeve, and as this would result in distortion if the piece were completely machined before hardening, we have to adopt the following procedure.

Cut off a piece of $1^1/_2$" b.m.s. to $3^3/_4$" long and get running true to gauge in the four-jaw chuck. Turn to 1" for $1^1/_8$" long and from $1^1/_2$" from the end up to the chuck jaws turn to $1^1/_4$" dia. Set up the indexing attachment for 30 divisions and the milling attachment for 20 D.P. and mill 30 teeth to a depth of 0.058" only. This is the dedendum of 20 D.P. and as the 1.50" dia. of the steel is the pitch circle diameter of 20 D.P. we will have the bottom half of the teeth of a gear which will form an excellent clutch. Now remove from the lathe and case harden thoroughly, but do not quench out, that is carburise only, the teeth of the clutch and the journal.

Set back in the chuck in the same position, running true to dial and turn $^1/_{16}$" off the end and turn the previously turned part up to the chuck jaws to $1^3/_{16}$". This will remove the carburising from these surfaces and the piece can now be heated to bright red and quenched in water. I forgot to say that the 1" × 20 t.p.i. on the end of the journal to a length of $^1/_4$" should also be screw-cut before quenching out.

Set back once again in the four-jaw chuck to exactly the same place and to dial gauge again to run true on the journal and clutch teeth and centre the end and drill and bore to 0.745" for 2" length and drill and bore to 0.620" for the balance of the length. This can now be machine reamed to $^5/_8$" right through, as the reamer will be long enough and also the outer end of the bore can be reamed $^3/_4$" for the full 2" length.

A proper mandrel should now be prepared to fit both the $^3/_4$" and the $^5/_8$" reamed holes to make a good true running job and the piece can now be turned to the drawing and the total length is faced up to $3^5/_8$" and the 1" × 20 t.p.i. screw-cut to $1^1/_{16}$" long. Two $^1/_4$" Allen grub-screws are fitted at 90 deg. at $^1/_4$" from the shoulder to hold the collets.

The phosphor-bronze bush is also $1^7/_8$" long, 0.752" o.d. and 0.626" reamed hole. This is pressed in and the reamer run through again as it will close in. This completes the main part of the large tapper and the nut I made from a piece of $1^3/_8$" A/F hexagon steel which I happened to have in the scrap box, but this again could be made from round b.m.s. with a good solid tommy bar to tighten it up. The tommy bar should be $^3/_8$" at least, which is why I prefer the hexagon rod. The piece is drilled right through $^{13}/_{32}$" and bored to $1^1/_8$" length to $^{15}/_{16}$" dia. and screw-cut 20 t.p.i. to a nice fit on the main body, and taper bored 30 deg. the same as the smaller nut.

The retaining collar $1^3/_8$" dia. and $^1/_4$" wide can also be bored to $^{15}/_{16}$" and also screw-cut 20 t.p.i., to fit its thread on the other end of the body. This part is fitted with a 4BA Allen grub-screw.

The nut is finished off outside to clean off the hexagon, leaving $^1/_2$" hexagon and to a total length of $1^3/_4$", turning off at 30 deg. leaving the end $^{11}/_{16}$" dia.

The fulcrum sleeve is of mild steel and is also case hardened all over after finishing. It is bored a free fit on the 1" dia. case-hardened journal, and is spherically turned outside to $1^1/_2$" dia., the centre of the spherical turning being set at $^5/_{16}$" from one end, leaving the opposite end at $1^1/_4$" dia.

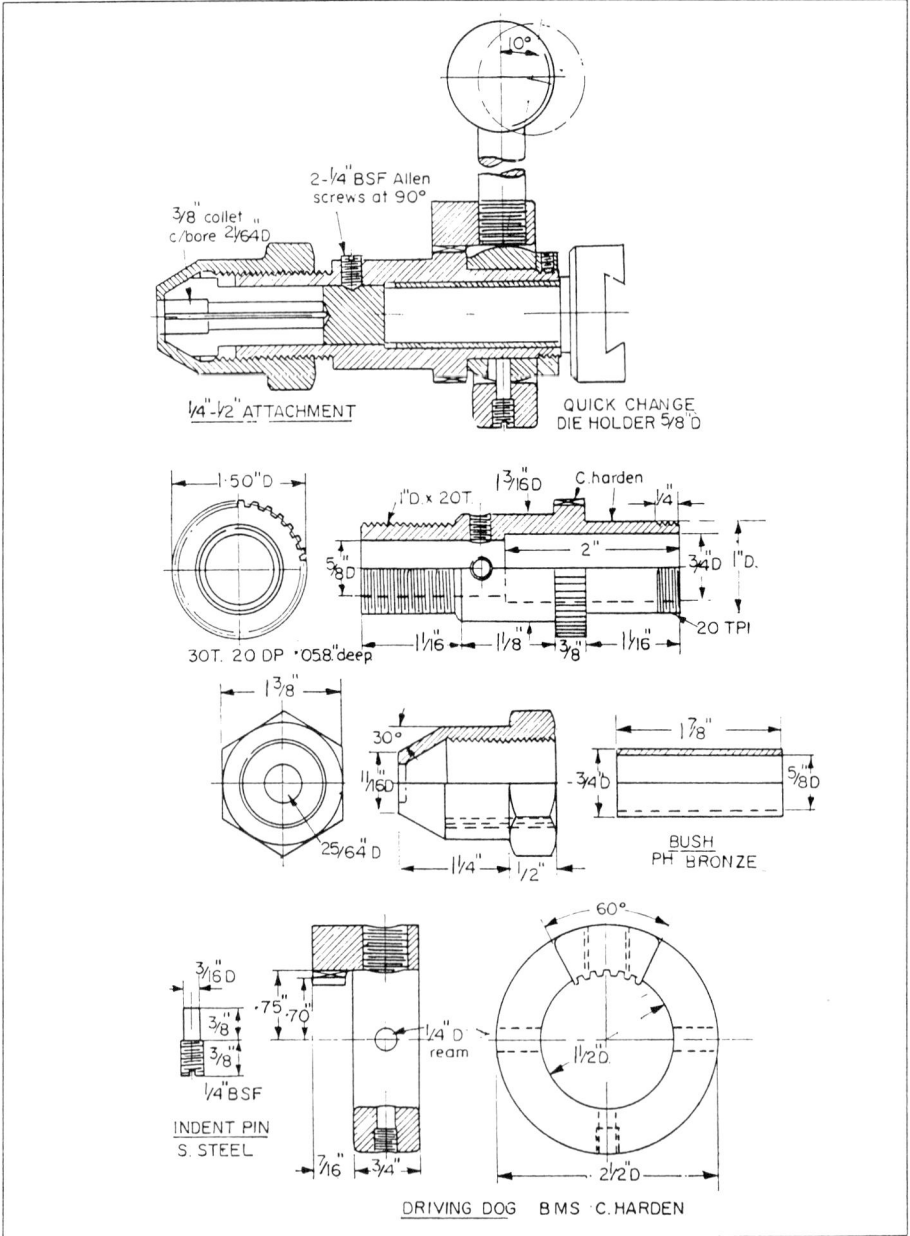

10°

2-¼" BSF Allen
screws at 90°

3/8 collet
c/bore 21/64 D

¼"-½" ATTACHMENT

QUICK CHANGE
DIE HOLDER 5/8 D

1·50"D

30T. 20 DP ·058" deep

1"D. x 20T.

1 3/16 D   C. harden   ¼

2"

5/8 D

3/4 D   1"D.

20 TPI

1 1/16   1 1/8   3/8   1 1/16

1 3/8

30°

1 1/16 D

25/64 D

1¼   ½

1 7/8

3/4 D

5/8 D

BUSH
PH. BRONZE

3/16 D

3/8

3/8

¼" BSF

INDENT PIN
S. STEEL

·75   ·70

¼" D
ream

7/16   3/4

60°

1½ D

2½ D

DRIVING DOG   B MS C. HARDEN

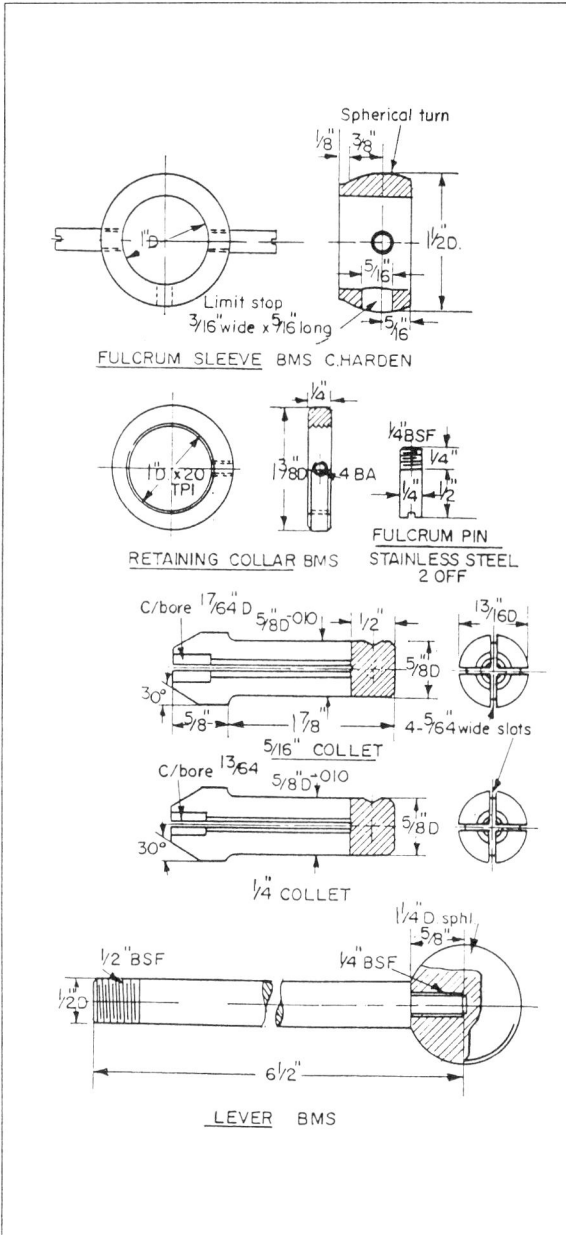

Spherical turn

FULCRUM SLEEVE BMS C.HARDEN

Limit stop
$\frac{3}{16}$" wide x $\frac{5}{16}$" long

RETAINING COLLAR BMS

FULCRUM PIN
STAINLESS STEEL
2 OFF

$\frac{5}{16}$" COLLET

$\frac{1}{4}$ COLLET

LEVER BMS

Two $\frac{1}{4}$" BSF holes are tapped across the diameter at $\frac{5}{16}$" from the end, bringing the holes to the centre of the spherical turning and at right angles to these two holes another hole $\frac{3}{16}$" wide, but elongated to $\frac{5}{16}$" long, is drilled and filed up to fit the limit pin and sufficiently long to clear the clutch teeth on the driving dog, when this is made and fitted up. Do not therefore harden out until this slot is checked with the driving dog. The two fulcrum pins and the limit pin can be made of silver steel and the fulcrum pins fitted temporarily.

The driving dog, a piece of $2\frac{1}{2}$" round b.m.s. $1\frac{1}{4}$" long, is held in the four-jaw chuck to run true, faced off and bored right through to 1.40" and then counterbored to 1.505" for a length of $\frac{7}{8}$". Chamfer the outer edge of the hole. Then with an internal pointed tool such as a screw-cutting boring tool, mark a ring inside the hole at $\frac{3}{4}$" from the outer edge. Reverse in the chuck and get running true. Mark two radial lines 60 deg. or one-sixth of the circle with a lathe tool, and now proceed to remove the unwanted metal with a hack-saw, sawing along the two radial lines down to your circle marked inside the hole and, after marking a circle on the outside of the piece, saw down to remove all but the 60 deg. segment as per drawing.

This segment is now slotted for 30 teeth to a depth of 0.058" using the indexing attachment and the slotting attachment (Chapter 6) with a shaped tool which should leave teeth which will fit closely the mating teeth, but will not quite bottom; this will need to be found by trial and error.

The sawn surfaces can now be dressed up with a smooth file and the two fulcrum pin holes drilled and reamed $1/4$" to fit the fulcrum pins, these are in the centre of the $3/4$" wide part and are truly radial and also the $1/2$" tapped BSF hole for the lever is drilled and tapped on the same line, but at 90 deg. to the other two holes and in the centre of the segment part.

The limit pin opposite the lever hole is now fitted. This is drilled first $3/16$" and then redrilled half way through $7/32$" and tapped $1/4$" BSF so that the pin can be securely tightened up without bottoming on the 1" journal of the main tapping part. The lever is $6^1/2$" of $1/2$" b.m.s. with a $1/2$" BSF thread to fit the driving dog and reduced to fit a $1^1/4$" spherical knob, which could be turned in place using the spherical attachment (Chapter 5) and afterwards blued in oil. After fitting up to try out the movement of the lever and clearance of the dog teeth, both the driving dog and the fulcrum sleeve should be case hardened and quenched in water.

The collets

There are two collets for the smaller tapper and three for the larger one. These collets grip on the round shank of the tap as well as on the square, thus holding it truly in line. They are milled with four slots into which the four corners of the tap fit. These are a little difficult to hold for slitting, so I have devised the following procedure for machining these collets:

Cut off two pieces of $1/2$" oil hardening non-shrink tool steel to 2" long and five pieces of $7/8$" or 1" to $3^1/8$" long. These pieces are longer than the finished collets will be, but it is necessary to leave a solid sleeve part on the tapered end of the collets unslotted, to hold the collets firmly together until all of the slits are milled.

Holding in the three-jaw chuck and with the two smaller pieces protruding $1^3/8$" and the three larger pieces protruding 2", face the end and centre. Turn the smaller ones to $3/8$"

a close fit in the smaller tapper and for $1^5/16$" long and undercut a further 0.010" from $1/2$" from the end to the end of the turning. This is to allow the collet to expand larger than its nominal size in the holder. The three larger pieces are turned to $5/8$" to fit the larger tapper and undercut the same amount, the length of the turning for these is $1^7/8$", and the cut should finish off with a nice radius.

Each piece should now be gripped by the last $1/2$" in the collet or running true in the four-jaw chuck and the fixed steady should be fitted close to the end of the turning as the collets must run true for the whole length for boring. Commencing with the largest collet, turn a parallel part up to $5/8$" from the previous turned shoulder to $7/16$" dia., and turn up to the shoulder $13/16$" dia. and setting the topslide at 30 deg. turn down until the tool touches the $7/16$" diameter part. This is now centred and drilled $21/64$" for a length of 2" and then drilled and reamed $3/8$" for a depth that will leave $3/8$" in the collet after the outer $7/16$" dia. end is cut off after slotting has been done.

The diameter of the extra end of the $5/16$" collet is $3/8$" and drilling to 2" depth is $17/64$" and the length of the $5/16$" hole in the collet is $3/8$". The $1/4$" collet is drilled $13/64$" and the extra length is $5/16$" dia. For the two smaller collets, the outer diameter is turned to $1/2$" to clean up and the diameter of the extra piece on the end of the $3/16$" collet is $1/4$", while the diameter of the smallest $1/8$" collet is $3/16$". Drill the $3/16$" collet $5/32$" for a depth of $1^1/8$" only and the $1/8$" collet $7/64$" for the same depth. The length of the nominal size in these two collets drilled through the extra end piece should leave $1/4$" of the nominal size in the collet after the end is cut off.

These extra ends are only cut off after the slitting has been done. Each collet is held next to the chuck in a piece of, say, 1" dia. steel, bored to a tight fit $5/8$" or $3/8$" with an Allen grub-screw to hold it. The outer end is held over a piece of turned and shouldered $1/2$" steel held in the drill chuck, of sufficient length to clear the slitting saw, and turned a close fit for each collet in turn.

Slitting the collets

The slitting is done by feeding downwards until the saw penetrates the centre hole in each case, but does not cut into the extra sleeve end on each collet, and it is traversed along up to the chucked end to within the $1/2$" long end piece which is held in the bored steel in the three-jaw chuck. The saw for the larger ones is $5/64$" thick and for the two small collets it should be $1/16$" thick. The smaller the diameter of the saws that can be used the better.

After all the collets are slit, the extra ends are sawn off by hand and the end of the collets are dressed up flat on the side of the bench grinder, turning around by hand. The saw cuts will then need to be finished into the tapered end and also straight across at the plain end with the hacksaw with two blades side by side and finished with a flat needle file.

Hardening and tempering

Before hardening, prepare a long tapered piece of $1/2$" steel tapered to a long point. As each collet is brought to a red heat, lightly push the tapered steel piece into the end to slightly spread the collet to above normal size to make it easy to insert the tap. Quench in oil and polish up for tempering. The tempering is done from the solid end and should be blued at the solid end and right along the slits up to the tapered end, which should be brought to a dark straw only and then the temper is stopped by quenching again in oil.

Before hardening, however, each collet should be tried out for fitting the various taps, as a little judicious filing in the slots can make a perfect fit so that the taps are held securely on the round of the shank as well as gripped on the squared ends.

This completes these two tapping attachments and I would strongly recommend model engineers to make them, as it saves a lot of wear and tear on the tailstock and drill chucks as well as on the operator.

CHAPTER 11

A Six-Position Saddle Stop

On many occasions when making a few identical parts, either between centres or in the chuck, I have felt the need for a multiple saddle stop, so that the parts would be accurate to length between the shoulders. Using a stop clamped to the bed is hazardous in the extreme, as should one be a little late in opening the leadscrew nut, the tumbler gears will be stripped or other damage will be done.

A saddle stop fitted to a lathe with a separate feed shaft, where the feed is operated by the rack, merely has to trip a latch and the feed drops out immediately; the saddle is then brought against a dead stop, ail of which is done in a very short length of travel. With the Myford, however,

the automatic feed is taken by the leadscrew and were it possible to operate the cam directly by the saddle movement it would operate only over a very long travel. However the cam operates against the direction of travel and has to be lifted to open the feed screw nut.

In my opinion, the operating cam should be lifted to close and knocked down to open; however it is possible to take advantage of the cam as it is and with my design the movement of the saddle, from the moment of contact with the adjustable stops to contact with the dead stop, takes approximately $1/8$", which is quite reasonable. An advantage also, is that the stops can be used for screwcutting, as when

The lathe set-up with the saddle stops in use, ready for a small repetition job.

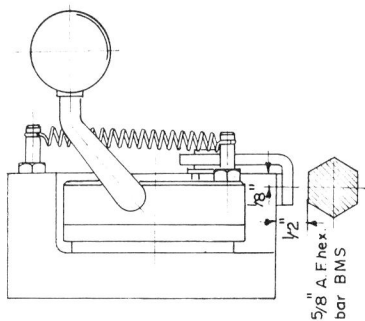

5/8" A F hex.
bar BMS

1/8"

1/2"

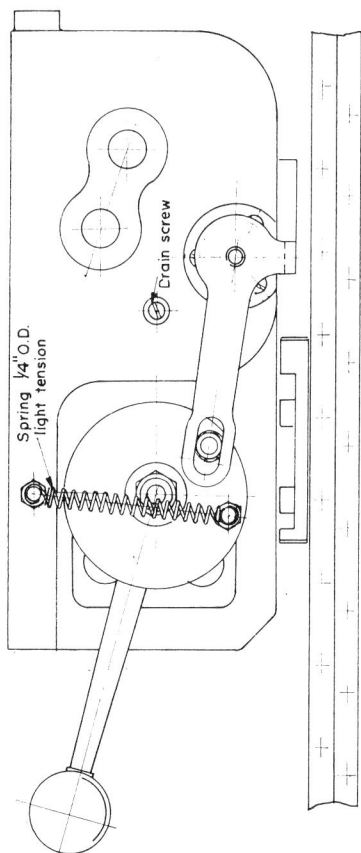

Spring 1/4" O.D.
light tension

Grain screw

SADDLE AND STOP BAR FEED LEVER 'OFF'

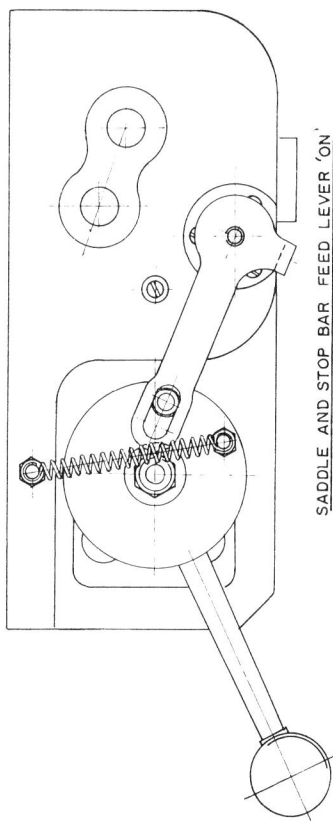

SADDLE AND STOP BAR FEED LEVER 'ON'

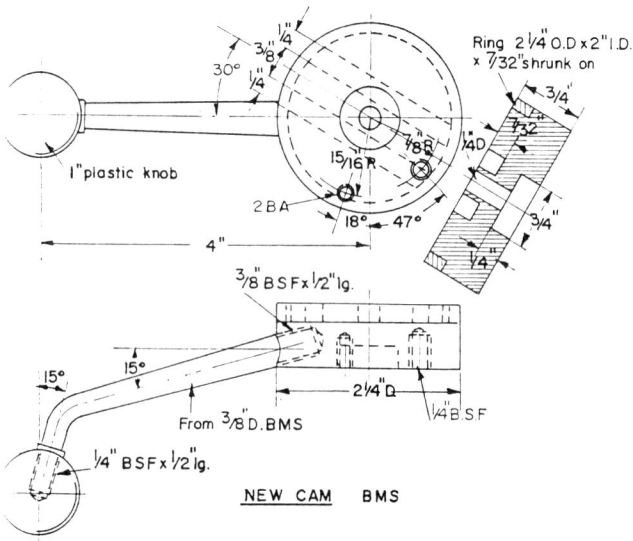

Ring 2 1/4" O.D x 2" I.D.
x 7/32" shrunk on

1" plastic knob

2 B A

NEW CAM   BMS

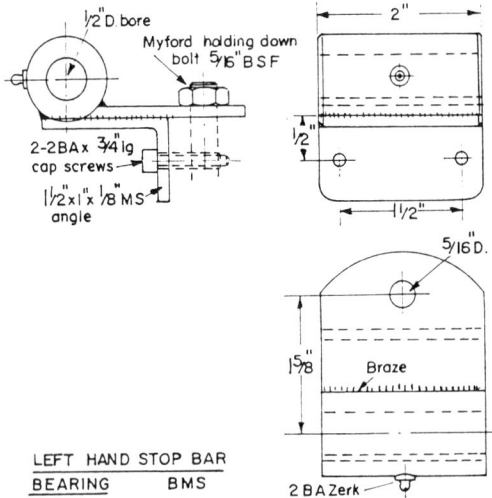

1/2" D. bore
Myford holding down
bolt 5/16" BSF

2-2 BA x 3/4" lg
cap screws

1/2" x 1" x 1/8 MS
angle

5/16 D.

Braze

LEFT HAND STOP BAR
BEARING   BMS

2 BA Zerk

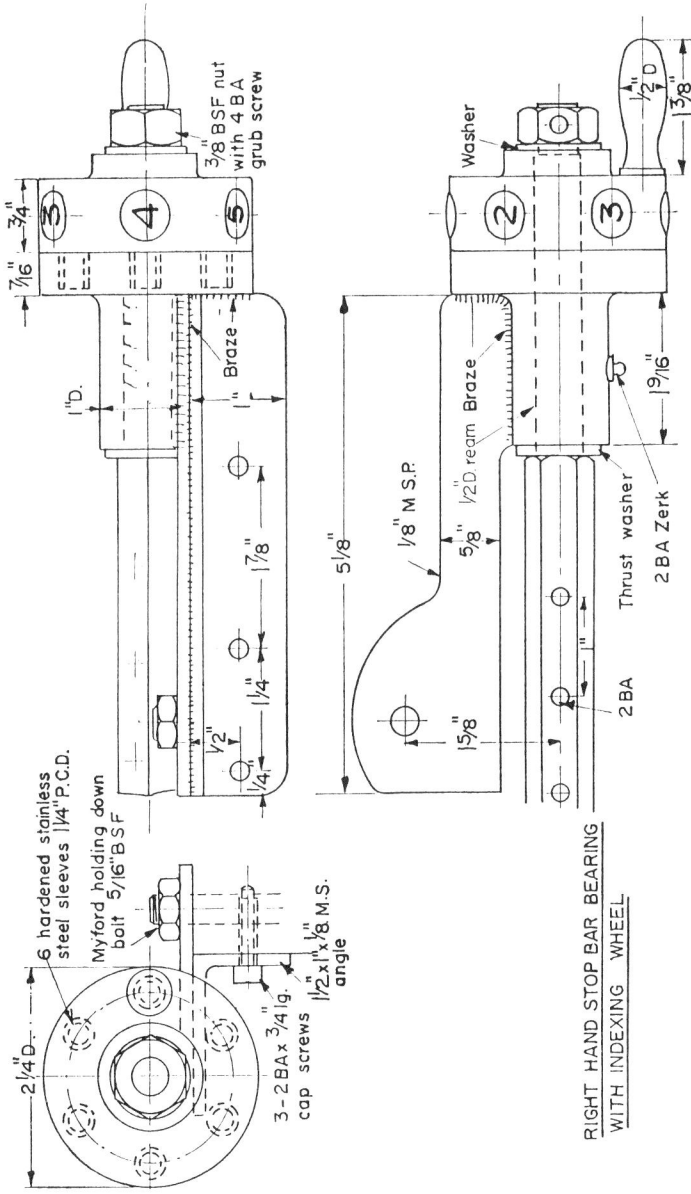

3/8" BSF nut with 4 BA grub screw

3/4"

7/16"

Braze

1" D.

1"

1 7/8"

1 1/4"

1/2"

1/4"

6 hardened stainless steel sleeves 1/4" P.C.D.

Myford holding down bolt 5/16" BSF

2 1/4" D.

3-2 BA x 3/4" lg. cap screws

1/2" x 1" x 1/8" M.S. angle

Washer

1/2" D

1 3/8"

9/16"

1/8" M S.P.

5/8"

5 5/8"

1/2 D. ream Braze

Thrust washer

2 BA Zerk

2 BA

1 5/8"

1"

RIGHT HAND STOP BAR BEARING WITH INDEXING WHEEL

105

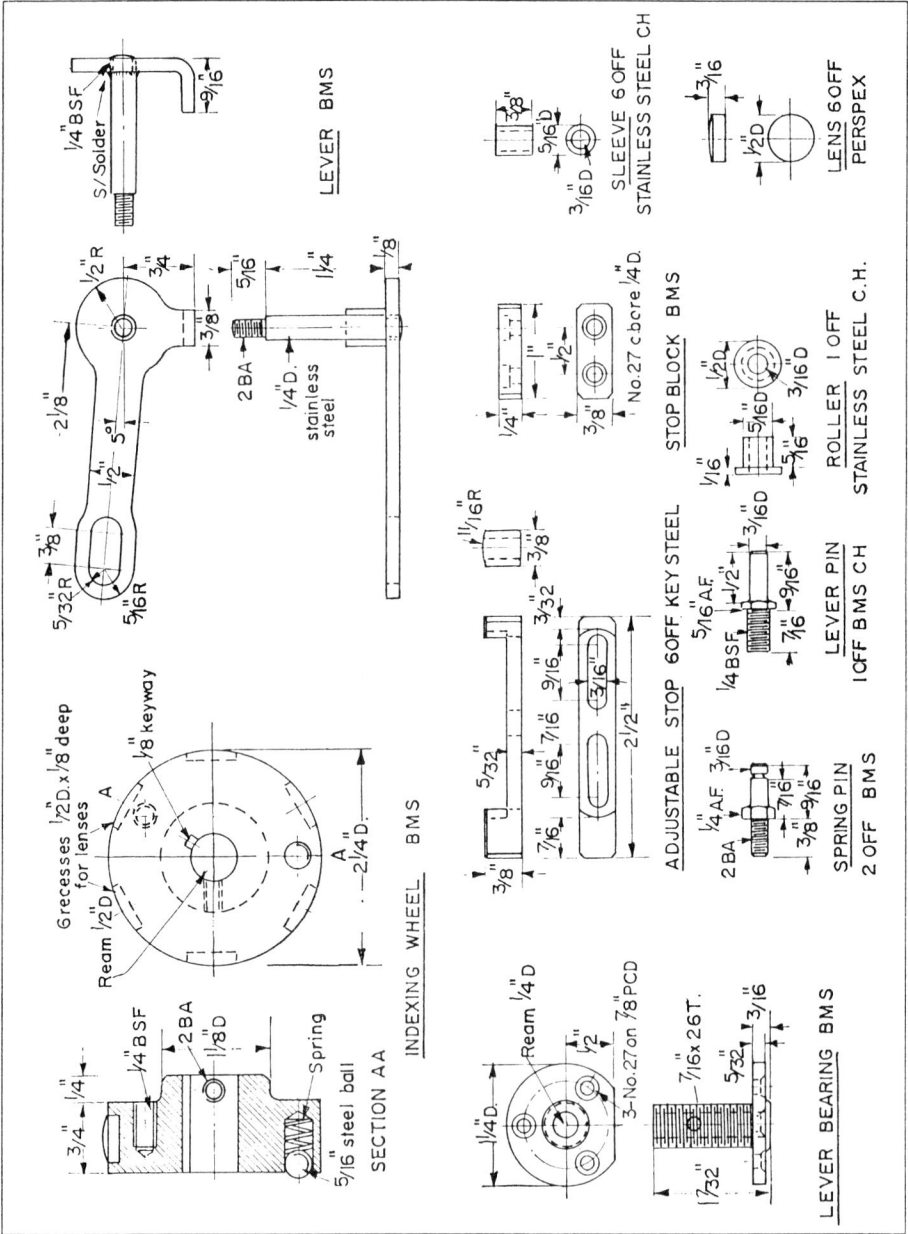

LEVER BMS

$\frac{1}{4}$ BSF
S/Solder
$\frac{9}{16}''$

SLEEVE 6 OFF
STAINLESS STEEL CH
$\frac{3}{8}''$
$\frac{5}{16}$ D
$\frac{3}{16}$ D

LENS 6 OFF
PERSPEX
$\frac{3}{16}''$
$\frac{1}{2}$ D

$\frac{1}{2}$ R
$\frac{3}{4}''$
$\frac{5}{16}''$
$\frac{1}{4}''$
$\frac{1}{8}''$
$\frac{3}{8}''$
2 BA
$\frac{1}{4}$ D.
stainless steel
$2\frac{1}{8}''$
$5°$
$5''$
$\frac{1}{2}''$
$\frac{3}{8}''$
$\frac{5}{32}$ R
$\frac{5}{16}$ R

STOP BLOCK BMS
$\frac{1}{4}''$
$\frac{3}{8}''$
$\frac{1}{2}''$
$1''$
No.27 c.bore $\frac{1}{4}$ D.

ROLLER 1 OFF
STAINLESS STEEL C.H.
$\frac{1}{2}$ D.
$\frac{3}{16}$ D
$\frac{5}{16}$ D
$\frac{1}{16}''$
$\frac{5}{16}''$
$\frac{1}{16}''$

ADJUSTABLE STOP 6 OFF KEY STEEL
$1\frac{1}{16}$ R
$\frac{3}{8}''$
$\frac{3}{32}''$
$\frac{9}{16}''$
$\frac{7}{16}''$
$\frac{9}{16}''$
$\frac{5}{32}''$
$\frac{7}{16}''$
$\frac{3}{8}''$
$\frac{3}{16}''$
$2\frac{1}{2}''$

LEVER PIN 1 OFF BMS CH
$\frac{5}{16}$ A.F.
$\frac{1}{2}''$
$\frac{3}{16}''$
$\frac{1}{4}$ BSF
$\frac{9}{16}''$
$\frac{7}{16}''$

SPRING PIN 2 OFF BMS
$\frac{1}{4}$ A.F.
$\frac{3}{16}$ D
$\frac{7}{16}''$
$\frac{9}{16}''$
$\frac{3}{8}''$
2 BA

INDEXING WHEEL BMS
6 recesses $\frac{1}{2}$ D. x $\frac{1}{8}$ deep for lenses
$\frac{1}{8}$ keyway
Ream $\frac{1}{2}$ D.
A
A
$2\frac{1}{4}$ D.

SECTION AA
$\frac{1}{4}$ BSF
2 BA
$\frac{1}{8}$ D
Spring
$\frac{3}{4}''$
$\frac{1}{4}''$
$\frac{5}{16}$ steel ball

LEVER BEARING BMS
Ream $\frac{1}{4}$ D
$\frac{1}{2}''$
3-No.27 on $\frac{7}{8}$ PCD
$\frac{1}{4}$ D.
$\frac{7}{16}$ x 26 T.
$\frac{5}{32}''$
$\frac{3}{16}''$
$1\frac{7}{32}''$

A close-up of the mechanism and stops.

screwcutting up to a shoulder or where it is necessary to finish the thread at a definite place, internally or externally.

There are six stops, all adjustable for the full length of the normal feed of the saddle; the feed stop applies, however, only to the normal 'towards the headstock' direction of feed, although it is possible to use the dead stop only in the direction of the tailstock if this should be required, such as, for instance, when making journals of a definite width.

The Myford cam and hand lever as fitted to the Super Seven is a one-piece casting and is not very elegant in appearance or convenient in operation. It is too small in diameter to use in this case so the first thing to do is to make a new cam and new lever with a plastic ball 1" dia. The ML7 has this type of lever, I understand, though I doubt if this would be large enough also. This cam, by the way, is retained in the open position by friction of a spring washer retained in adjustment by a Simmons nut. The new cam dispenses with the spring washer, but retains the little plain washer and the Simmons nut and is held in the open and the closed position by a light tension spring and is otherwise perfectly free in operation, making a much more sensitive feel to the engaging of the lever and a snappier action. The centre of the knob is also set out from the face of the cam $1^3/_4$", away from the hot swarf.

About $2^1/_2$" of $2^1/_4$" dia. bright mild steel will be required for the cam and the shrunk on ring which keeps out swarf; it is set up in the 4-jaw chuck, faced, centred and drilled letter D and reamed $^1/_4$" for about $1^1/_4$" depth. There are two $^1/_4$" wide by $^3/_{16}$" deep grooves milled across the face at $^5/_8$" centres. These dimensions should be checked from your own cam and rigidly adhered to. The two grooves must be concentric with the centre hole also, or one half of the leadscrew nut will close before the other, bending the leadscrew. Set up a $^3/_{16}$" wide slitting saw for milling and mill as close to $^5/_8$" centres as can be measured. After milling to a full $^3/_{16}$" or $^7/_{32}$", measure both the width of the grooves and the centres and correct accordingly by means of the vertical slide and then run through again until the slots are 0.250" wide and 0.625" centres, and truly concentric with the centre hole. Turn the outer edge of the cam to 2" dia. to a depth the same as the bottom of the slots; this is to take a shrunk-on ring to keep out swarf, as I have noticed that each time I have removed the original cam, there has been quite a lot of steel swarf jammed in the grooves.

Part off at about $^{13}/_{16}$" thick so that the cam can be finished to $^3/_4$" thickness in the 3-jaw chuck, but before this is done, bore the ring 0.003" smaller than the turned shoulder and part off to the correct width so that both faces will come flush. You will, of course, have taken a very light skim over the outside of both the cam and the ring at the same time. The outer face of the cam can now be faced off to $3/4$" to a good finish and the outer edge slightly rounded. Shrink on the ring and finish off the face on a sheet of emery cloth on a flat face, so that this working face is as friction free as possible. You will notice that the handle on the original cam is set at 30 deg. to the slots. File a flat on the diameter of the earn at this position and at 15 deg. cutwards from the face and centre drill at these angles $3/16$" from the front edge; drill and tap $3/8$" BSF to $1/2$" depth only; do not break into the slots.

The handle, a piece of $3/8$" b.m.s., $4^1/4$" long, is screw-cut $3/8$" BSF for $1/2$" long, turned taper to within $9/16$" of the other end down to $1/4$", leaving a little ring for appearance and providing a shoulder; it is then turned and threaded $1/4$" BSF for the 1" plastic knob $1/2$" long. This is screwed tightly into the cam and can be pinned if necessary and then is heated to redness about $1/2$"

from the little shoulder and bent so that it stands outwards within 15 deg. of the right-angle.

The two $1/4$" pins which screw into the leadscrew nut halves were slightly worn in my case, so in order to reduce all lost motion to the minimum, I made two new ones from $5/16$" silver steel, turning down and screwing to a close fit in the new cam. These were made the maximum possible length to clear the bottoms of the grooves and were hardened dead hard, tempering only the threads.

In order to find the correct positions for the spring pin and the roller pin, the cam can now be fitted into place. Assemble the cam, but without the spring washer, and locate the maximum 'closed' position by unscrewing the adjusting screw fitted through the lower half of the leadscrew nut and closing tightly around the leadscrew. Adjust the screw until the leadscrew is just free or until there is no bottoming friction. In this position, find the vertical centre below the centre shaft using a square underneath the cross-slide. Mark a line vertically below

The modified saddle and some of the parts.

The inside of the saddle showing the oil feed pipe.

the centre shaft and, at $3/16$" to the right of this line and at $15/16$" radius, centre punch lightly. This is the position of the spring pin and the spring retains the cam both in the open and in the closed positions. The top pin is marked on the vertical centre-line and in the centre of the top face of the apron.

The pin for the lever roller is found by first marking off the position for the lever pin. At $1^1/2$" from the edge of the middle part of the apron ($2^3/4$" from the right-hand edge), and at $1/2$" from the bottom face, lightly centre punch for the lever pin and now, with the cam lever closed, mark a straight line between this centre and the centre of the cam. At $7/8$" radius, lightly centre punch the position for the lever roller pin. The beginning of the movement is then on the radial line and will give the maximum torque at starting. Drill and tap 2BA for the two spring pins and $1/4$" BSF for the lever roller pin. Do not drill into the cam slots.

The two spring pins are made from $1/4$" a.f. hexagon b.m.s. and the lever roller pin from $5/16$" a.f. hex. b.m.s. This one is case hardened on the $3/16$" plain part for the roller.

The roller is made from $1/2$" dia. silver steel and is turned $5/16$" dia. for $1/4$" long, drilled and reamed $3/16$" for the pin and parted off at $5/16$" long which will leave a retaining flange $1/16$" thick. This is hardened right out in oil. The spring is $1/4$" o.d. of 0.030" piano wire and need not be of greater tension than will retain the cam definitely in the open position: the lighter the better. This new cam can now be fitted and used so that one can get accustomed to the new improved action.

The lever bearing is made from $1^1/4$" b.m.s. and is turned down and screw-cut $7/16$" × 26 t.p.i. for a length of $1^1/16$", drilled letter D and reamed $1/4$". The flange is $5/32$" thick increased to $3/16$" in the centre part at about $1/2$" dia. The three 4BA countersunk screws are not drilled for, nor is the side cut off until the part is fitted permanently into place.

The lever is made from $1/8$" b.m.s. plate and the two arms are at 95 deg. The centre

is tapped $1/4$" BSF and the silver steel spindle is tightly screwed into place and is then silver soldered. I dull chromed this lever after finishing to hide evidences of brazing. The lever is made with slight end clearance in the lever bearing and is fitted with a small washer and a 2BA nut. The slot for the roller is at $2^1/8$" from the centre and is $3/8$" long. These distances should be checked for clearances each end of the movement however, and when found correct, the end is case hardened for the roller to engage against. The vertical arm is bent underneath at $3/4$" to the outside from the centre of the spindle and is $9/16$" long to the outside. The arm itself is $3/8$" wide.

The apron can now be removed from the lathe. It will be necessary to remove the leadscrew handwheel and bearing from the right-hand end of the bed in order to refit the completely assembled and adjusted apron, so it may as well be removed now to make removal easier. The saddle handwheel is removed, the gear case from inside the apron and all the gears. The back of the apron is machined flat so that it can be clamped to the drilling machine table and the hole for the lever bearing is centre drilled and drilled to fit a spot facing cutter and the face just cleaned up to a clearance on $1^1/4$" dia. It is then drilled $25/64$" and tapped $7/16$" × 26 t.p.i.

I spread a little Araldite on the bearing flange and the face machined on the apron and screwed the bearing tightly home. The three 4BA countersunk screws were fitted and the flange sawn off flush with the bottom of the apron. I then drilled a hole from the inside of the gear case on the inside of the apron to fit a $5/32$" copper tube, this hole being carried through into the bearing flange. A copper tube was fitted to run oil down from the gear case into the bearing flange, the top of the copper tube being left about $1/8$" high and so retaining oil in the gear case. I also drilled a $1/8$" hole at the lowest point on a downward slope for a drain hole until the drill just began to show through on the front of the apron. This was then drilled and tapped 2BA and a large flat headed 2BA brass screw was fitted as a drain screw. The reason I fitted the drain plug was that the gear case inside the apron was full of Solvac and water which had displaced the oil. I can now drain the gear case periodically and replenish the oil.

The stop block, made from a 1" length of $3/8$" × $1/4$" key steel is fitted with two 4BA cap screws set in flush with the surface. It is fitted flush with the front of the apron and the position longitudinally is determined by the position of the lever when the leadscrew nuts are sufficiently clear of the leadscrew, although not necessarily right up to the old position which was determined by the length of the cam pin slots. The lever should just clear and perhaps a further $1/8$" on the plastic knob will give sufficient clearance. The lower face of the stop block is rounded off and the two ends bevelled. The apron was now thoroughly washed out and re-assembled, the leadscrew nuts adjusted to work freely but without play, the lever fitted and nutted up, and everything worked perfectly freely. The apron was passed over the end of the leadscrew and re-assembled on the saddle.

For the stop bar, a length of $5/8$" a.f. hexagon free-cutting mild steel is required. Do not accept 0.600" a.f., as this will be too narrow on the flats for the adjustable stops. An extra 1" each end is required as it must be turned in the chuck and with a sleeve over the hexagon, running in the fixed steady, as it will be too long to go between centres. For my long bed lathe, 3 ft. 6 in. was required; for the short bed lathe the length will be 10" longer than the distance between the two $5/16$" Myford holding down bolts which go through the feet of the bed. A sleeve is bored to fit closely over the corners of the hexagon and turned true outside and about 1" long, which will allow the fixed steady to be used. The steady is fixed outside the saddle and turning is done close to the chuck.

It will be better, however, to make the stop bar bearings first and fit them into

place so that the exact distance between shoulders can be easily determined. The right-hand bearing is made from a piece of $2^{1}/4$" b.m.s. and is turned down to 1" dia. for $1^{9}/16$" leaving a good radius. It is then parted off at $2^{1}/16$". The 1" dia. can now be held in the chuck and the end faced, leaving the full dia. at $7/16$" long. It is then centre drilled and drilled $15/32$", lightly bored to true up and reamed $1/2$". While still set up, the indexing head is set up and six positions are drilled with a small drill through a drilling jig held under the toolpost, the radial position from the centre being $7/8$". Supporting the bearing face upwards on the drilling table on blocks or on the jaws of a vice, these six holes are drilled $19/64$" to $3/8$" depth to the point of the drill only, then drilled $5/16$" and the bottoms of the holes drilled flat to $3/8$" depth using a flat bottomed drill or D-bit, the depth being determined by a stop on the drilling machine. These six holes are to take hardened steel sleeves into which the steel ball in the indexing wheel runs. The left-hand stop bar bearing is a 2" piece of 1" dia. b.m.s. faced off each end and bored and reamed $1/2$" right through.

Indexing wheel

The indexing wheel is also of $2^{1}/4$" dia. b.m.s. faced off, drilled and reamed $1/2$" and a keyway cut $1/8$" wide and about $1/8$" deep. At 30 deg. ($1/12$ turn on indexing plate) from the keyway and at $7/8$" radius, a $5/16$" hole is drilled for the indexing steel ball $5/8$" deep to the point of the drill, using the drill jig again under the toolpost. Again with the drilling jig placed close to the outside diameter and on the same radial line as the keyway, six positions are drilled for the numbers at $3/8$" from the face. The piece is now parted off to just over 1" long and reversed in the chuck and turned to $1^{1}/8$" dia. for $1/4$" long, leaving $3/4$" at the full $2^{1}/4$" dia. The six indexed places around the outside are now drilled $1/2$" until the lip of the drill is entered and then drill to $1/8$" depth with a $1/2$" D bit, the $1/8$" depth being at the shallowest places.

The $1/4$" BSF tapping for the handle is in

line with the keyway and the hole should just clear the bottom of the recess under which it is drilled. A 2BA grub screw is fitted at 45 deg. angle in the corner of the $1^1/_8$" dia. shoulder, at right-angles to the keyway. The handle is turned from $^1/_2$" b.m.s. and is $1^3/_8$" long, plus $^1/_2$" for the $^1/_4$" BSF thread.

The six numbers are printed on white glazed card from 'Letraset' $^1/_4$" numbers obtainable from any drawing office supply shop and are stamped out afterwards using a $^1/_2$" gasket punch. The six lenses are turned from $^1/_8$" Perspex sheet between two pieces of $^1/_2$" b.m.s. rod, one held in the chuck with a tiny pin-point turned in the centre to grip the Perspex and the other running over the Allen screw in the smallest ball bearing cone centre. These could more easily be turned and parted off from $^1/_2$" Perspex rod if obtainable. The outside face is domed slightly and acts as a lens. It is polished with 400 paper dry and then polished with Perspex polish or Brasso. The six numbers are put in, the numbers running consecutively clockwise looking at the flat face of the index wheel and the lenses are secured with a little Perspex glue around the edges. I blued the index wheel in oil before fitting the numbers. The spring for the steel ball is quite strong and should be fitted before allowing the ball to enter the hole. A thrust washer $^7/_8$" o.d. by $^1/_2$" hole and $^3/_{32}$" thick is fitted over the stop rod and a $^1/_{16}$" thick washer $^7/_8$" o.d. by $^3/_8$" hole goes against the adjusting nut. This nut is made from a short end of the hex. steel and is tapped $^3/_8$" BSF and fitted with a 4BA grub screw to lock it.

These two stop bar bearings are brazed to brackets made from $^1/_8$" b.m.s. plate and $1^1/_2$" × 1" × $^1/_8$" b.m.s. angle. The $^1/_8$" plate is supported under the main holding down $^5/_{16}$" bolt which goes through the foot of the lathe and the angle is brazed to this and is supported on its vertical face by 2BA Allen cap screws tapped into the bottom edge of the lathe feet. The bearings are filed to a flat underneath and rest on the angle mild steel only, the $^1/_8$" plate butting against the inside

radius only. The amount to file off the bearing sleeves is important, as this determines the height of the stop bar. The top flat of the stop bar should be $^1/_2$" below the bottom surface of the apron with the saddle run to each end of the bed. Also, the lateral position of the stop bar is important. The centre of the bar should be $^1/_8$" behind the front face of the saddle. As can be seen in the right-hand end view of the saddle on the first drawing, the two adjacent stops are drawn to show this position and the clearances necessary.

It will be found easier to bolt the $^1/_8$" plate and $^1/_8$" angle together temporarily until the proper shapes and the height and position of the stop bar is determined. A piece of $^1/_2$" b.m.s. round can be passed through the bearings which are then brought into alignment so that the bar turns freely with the bearings tightly clamped to the angles and resting against the edge of the $^1/_8$" plate. The flat filed on the right-hand bearing is parallel to two of the index holes; this is also important and filing should be done to the micrometer. With the bearings clamped up tightly, the $^1/_2$" bar in place, height and position of bar correct, remove the bar, remove the bearing brackets with bearings still clamped and drill and tap two 2BA holes underneath into the bearings and fit two screws. Replace and test again to make sure that nothing has moved.

I had forgotten to mention that the three 2BA cap screws should be fitted to the right-hand bearing angle and the two ditto to the left-hand bearing angle before removing for drilling. If all is correct, the bearings can be brazed at all contact places and the edges of the plate and angle also brazed. Replace and test again with the $^1/_2$" rod and if distortion has taken place, it can be carefully set, using the $^1/_2$" rod as a lever.

All that remains now is the stop bar and the six adjustable stops. The stop bar is turned next, the 3-jaw chuck using the sleeve over the hexagon in the fixed steady on the other side of the saddle. Turn to 0.498" and to a smooth finish. The thread can be cut in this position, finishing with dies after the rod is parted off. A Woodruff keyway is cut $1/2$" × $1/8$" in position for the indexing wheel and this must be in line with one of the flats of the hexagon. The length of turning for the right-hand end is $3^1/8$" plus $5/8$" of thread ($3/8$" BSF) to allow for the $3/32$" thick thrust washer and to allow for thrust adjustment, and the length of the left-hand end is $2^1/8$" to allow $1/16$" clearance and $1/16$" to protrude through the end. One small point, a $1/4$" BSF zerk nipple for oil is set in the centre of each bearing and at 45 deg. to make for easy application of the oil gun. Paint the bearings and angles to match the lathe.

A jig is made for drilling the No. 24 holes in the stop bar. This is a piece of $1/4$" thick b.m.s. filed to exactly the same width as the hexagon rod is across the corners (mine is 0.713"), and two No. 24 holes are drilled at 1" centres right in the centre. A pin is driven into one hole and the other hole is used as a drilling jig. The hexagon rod and the jig above it are both clamped in the machine vice, using a piece of card under one jaw only. Drill to $9/16$" depth only, so that no burrs are raised on the underside, which upsets the level of the set-up. The first hole is drilled about $2^5/8$" from the left-hand end and the row of holes is carried to within about $3/8$" of the other end. The bar is now turned with the second flat uppermost and, at one third of an inch from the first hole in the first flat, a hole is drilled and the row of holes carried along as before. The third flat is now drilled $1/3$" away as before and continued right along.

Now drill all the holes right through the bar putting the bar over two pieces of key steel to clear the burrs, and lightly countersink both sides of the holes so that the tap will not raise a burr. All the holes are now tapped right through 2BA. I used a second tap through from one side and followed with a plug tap from the other side.

Twelve 2BA cap screws cut down to $7/16$" long will be required for the adjustable stops.

Six pieces of $3/8$" key steel $2^1/2$" long are drilled $3/16$" right along the centre with four holes. The first hole is drilled $1/4$" from one end, then $9/16$" centres, then $7/16$" centres and then again $9/16$" centres. These stop bars you will notice are offset, so that they can be reversed end for end and so any position along the stop bar can be used.

A $1/2$" end mill is now used and the thickness of the key steel is reduced to $5/32$" leaving $3/32$" at one end and $7/16$" at the other end of the original thickness; this is to accommodate the Allen cap screws and is carried $1/16$" beyond the edge of the two outer holes. A $3/16$" end mill is used now to mill out the space between the outer pairs of holes, leaving $1/4$" unmilled in the centre. Bevel the ends of the stops $1/16$" and radius off the tops of the stops to about $11/16$" radius.

When not in use, this stop bar in no way interferes with the hand operation of the lathe; in fact you will find the action of the cam lever is much improved. In use, I find it better to help the feed over the last $1/8$" with the hand wheel and so relieve the nuts of wear on a diminishing area as the nut opens. There is no need to touch the cam lever, however, and this leaves the left hand free to operate the cross-slide handwheel in the normal way instead of having to change over hands with the right-hand on the cross-slide wheel while the left operates the cam lever. It becomes simple to run out a thread when screwcutting without having to turn a groove first as when making a lathe mandrel nose thread. For internal boring and screw-cutting, all anxiety is removed. For setting the stops to that last thou, use the leadscrew handwheel, either by moving the stops with the saddle while slackened off slightly, or bringing the stops against the saddle after setting.

CHAPTER 12

A Graduating Tool for Cylindrical, Angular and Flat Surfaces

It could be truly stated that the difference between the professionally-built engineering article and some amateur work is not the result of lack of skill so much as the lack of facilities for producing that professional finish and accuracy. The amateur in many cases will make do with some makeshift set-up and produce a good job by sheer skill. The engineering firm would never do this the results would be too uncertain and costly. A special machine or fixture would always be designed and made to make the application of skill unnecessary.

A particular case in point is the graduating of collars on the cylindrical or angular surface and the graduating of flat surfaces in degrees for angular indexing. The usual way is to set a tool, ground to a vee shape on its side, under the toolpost of the lathe, and to use stops of different lengths mounted on the lathe bed and to operate against the saddle. Unless the tool is mounted absolutely on the centre height, particularly when graduating flat and angular surfaces, the faults are very noticeable; the lengths of the lines generally are not particularly regular as a rule, which again brands the work as amateur.

I have therefore designed a graduating tool which slips under the toolpost of the lathe at any angle without the need to alter the topslide, and which will graduate at any angle, cylindrical or flat, in a fraction of the time, as there is no setting up; it will produce that professional finish every time. The cast-iron body, which is held under the tool clamp of the lathe, has a ram sliding in vee ways, provided with a tool at each end, one for cylindrical work and the other for flat and angular work. The ram works under the action of a pinion, the ram having rack teeth cut along one face. The pinion is cut integral

The graduating tool.

A GRADUATING TOOL FOR CYLINDRICAL, ANGULAR & FLAT SURFACES

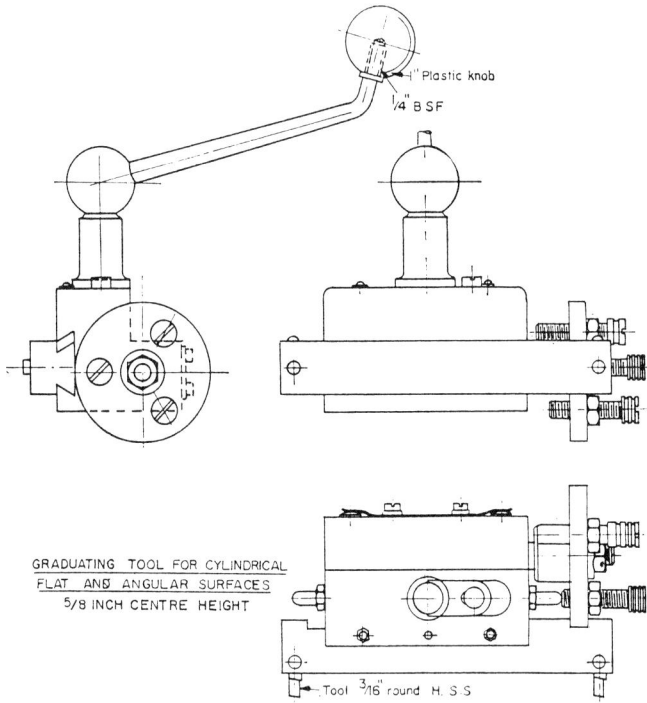

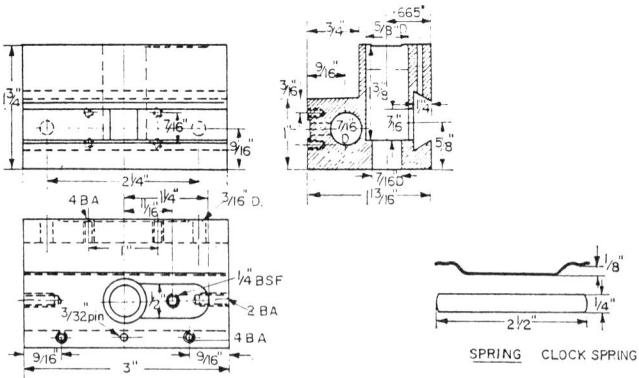

GRADUATING TOOL FOR CYLINDRICAL
FLAT AND ANGULAR SURFACES
5/8 INCH CENTRE HEIGHT

SPRING CLOCK SPRING

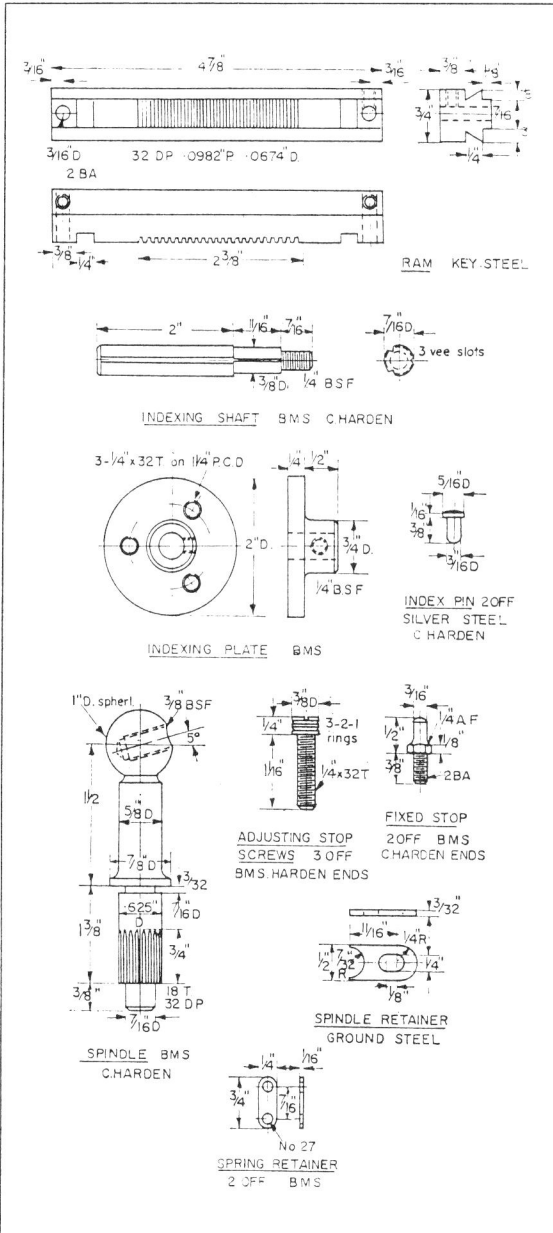

3/16"D   32 DP   ·0982"P   ·0674"D
2 BA

RAM   KEY·STEEL

INDEXING SHAFT   BMS   C·HARDEN

3 vee slots
3/8"D,   1/4 BSF

3-1/4"x32T on 1/4 P.C.D

INDEXING PLATE   BMS

INDEX PIN 2OFF
SILVER STEEL
C HARDEN

1"D. spherl   3/8 BSF

3-2-1
rings
1/4x32T

ADJUSTING STOP
SCREWS   3 OFF
BMS. HARDEN ENDS

FIXED STOP
2OFF   BMS
C HARDEN ENDS

SPINDLE RETAINER
GROUND STEEL

SPINDLE   BMS
C.HARDEN

No 27
SPRING RETAINER
2 OFF   BMS

with the operating shaft and handle. The spindle can be slipped into any operating position on the ram so that the handle can be brought well clear of the work or any other obstruction. The indexing plate is provided with three adjustable stop screws, providing three lengths of line between one half inch and zero. It can be slipped out of one end of the main body and slipped in the other end without the need for spanners. This allows for graduating to left or right or to or from the centre of a flat.

The work can all be done on the lathe and, as the slide for the ram and the drilling of the tool holes is done in position under the clamp of the topslide, the fixture is inherently true.

The pattern for the body could hardly be simpler. Allow a good 1/16" on each face and the faces can be left square as the foundry will set the pattern in the sand corner to corner so making the 'draw' 45 deg. The body can be faced on the four sides and the two ends as a face milling operation on the cross-slide of the lathe, using the four-jaw chuck. It could also be done on the vertical slide of the lathe, or in the shaper, and the 3/4" × 3/4" cut-away for the lathe tool clamp can be milled or shaped at the same time. It is 1 3/4" high, 1 13/16" wide and 3" long (finished sizes) with the corner milled away to leave 1" height under the tool clamp and 1 1/16" wide at the top.

Before boring for the operating spindle, make the spindle and the ram, so that the centres can be checked and boring done correctly. The spindle is made from 1" b.m.s. and is machined to

Set-up ready for boring for the pinion shaft.

the drawing. The pinion teeth are 32 d.p. and 18 teeth, the o.d. is 0.625" and depth 0.0674". The teeth were cut in the lathe using one of my hob-type cutters. This indexing fixture or something similar will of course be necessary for future graduating using this tool. The ball end was turned using the spherical turning fixture. It can be seen now that each fixture made makes other fixtures and tools possible.

The ram is made from 5" of $3/4$" key steel, and was machined using my 'Bormilathe' type elevating heads. My photograph shows the set-up for milling the rack using these heads. The ram of course can be milled in the lathe using the vertical slide and end mills. For those who have made my Quick Change Toolholders, they will have made or purchased a $7/8$" dia. dovetailed cutter and this cutter will be required for milling the ram and for milling the slideways in the main body. The rack teeth can be milled either by setting in a vice and packing up on the cross-slide to the correct height, or using a set-up on the vertical slide, parallel to the lathe centres, either in front of or behind the mandrel and using, in the absence of a rack cutter, a flycutter made to $14^{1}/_{2}$ deg. included angle, the flat tip of which is, as near as I can

measure it, 0.038-0.040" wide. Depth 0.0674".

Two $1/4$" wide grooves $1/8$" deep are milled, one at each end, to engage the index plate and they are $3/8$" from each end. Do not drill the tool holes. You will notice that the rack is $7/16$" wide and is offset from the centre $1/16$". The rack teeth could also be cut on the shaper. The spacing of the teeth is 0.0982" centres and the saddle stop is brought against the saddle, the saddle clamped and the first tooth cut; then a piece of steel of 0.0982" thickness is held between the saddle and the stop after moving the stop along, the stop is clamped and the saddle moved into contact with it again after removing the piece of steel, the saddle clamped and the next tooth cut, and so on for each tooth.

With the spindle and the rack machined, the two can now be run together to make the action smooth. Use some grinding paste of the valve grinding type and, with the pinion held down into tight contact with the rack by pressing downwards with a piece of wood, roll the two together until the action feels smooth. Case harden the pinion shaft all over; the rack should not be hardened or it will distort badly. A final roll along with fine paste after hardening will make a good job.

A GRADUATING TOOL FOR CYLINDRICAL, ANGULAR & FLAT SURFACES

The handle is turned taper from $^3/_8$" b.m.s. and is screwed $^3/_8$" BSF on the large end and $^1/_4$" BSF on the small end to fit the 1" dia. plastic knob. Screw the handle tightly into the spindle (the hole in the spindle, by the way, is drilled 15 deg. to the horizontal), and heat close to the small end and bend to within 15 deg. of the vertical.

It is now necessary to find the exact position for boring the hole in the main body for the spindle. Holding the rack and pinion into close mesh and with a feeler gauge between the $^7/_{16}$" dia. and the milled face of the ram, this distance can be measured exactly. This distance multiplied by two, plus the diameter of the $^7/_{16}$" end (by micrometer), plus twice the $^1/_4$" distance to the outer milled face, will give the diameter of a disc which, if brought flush with the edge of the body, will give the centre position for boring the hole for the spindle. Make a disc to press right on the female centre held in the headstock of the lathe and turn and face to this diameter plus 0.002" for clearance. Then with the main body held in the four-jaw chuck and in a central position, making sure it is square to the face of the chuck, bring the turned disc, held in the tailstock, into contact with the face of the body and the edge of the disc flush with the edge of the outer face of the body. This will give the exact position for boring for the spindle.

The body is now bored and reamed right through $^7/_{16}$" and bored $^5/_8$" and reamed to $1^3/_8$" depth. At this setting I also milled the slot for the spindle retainer to $^3/_{32}$" depth and $^1/_2$" wide to a length of $^7/_8$" from the edge of the $^5/_8$" hole.

The body should now be clamped to the vertical slide facing the chuck for milling the slide faces for the ram. Set the body square by resting it on a piece of key steel in one of the tee slots of the vertical slide. and set up a $^5/_8$" end mill in the chuck or the collet. The edge of the end mill should be $^3/_8$" from the bottom edge of the body; this can be assured by using a piece of $^3/_8$" key steel touching the end mill and flush with the bottom edge. Mill to 0.250" depth, setting by means of leadscrew handwheel graduations. Then set the $^7/_8$" dia. dovetail cutter in the chuck and at the same vertical setting, and to exactly 0.250" depth, mill right through also.

Now set a $^1/_2$" end mill in the chuck and with this set so that its edge is $^3/_8$" from the bottom edge in the same manner as was the $^5/_8$" end mill (this is not central as allowance must be made for the gib piece), mill to a further $^1/_8$" from the bottom of the previously milled slot, this is to clear the rack portion of the ram, and depth should be exact so that the ram can also bear on this face.

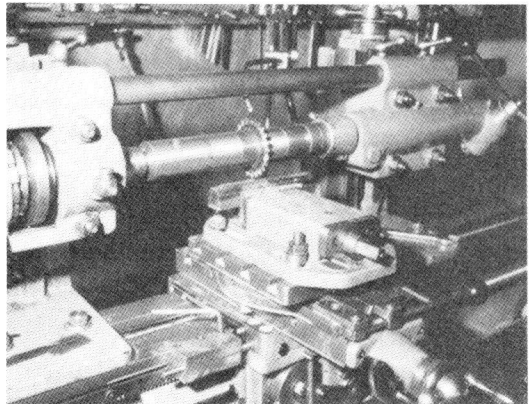

Milling the rack teeth using elevating heads and quartering table with vice (note saddle stop).

A GRADUATING TOOL FOR CYLINDRICAL, ANGULAR & FLAT SURFACES

Graduating tool set-up for face graduating.

If care has been taken with these measurements, the spindle and ram will work together with just perceptible play. The gib piece can now be made to a close fit and flush with the face of the body. A $^3/_{32}$" dowel pin is fitted in the centre through the gib to retain it and two 4BA gib adjusting screws with very small nuts are made and fitted. These are drilled at $^9/_{16}$" from each end. The two holes for the two $^3/_{16}$" dia. graduating tools can next be drilled and reamed. The hole for the tool that is used for graduating on flat surfaces will be simple to drill. Set the attachment under the toolpost of the lathe facing the chuck and, at $^3/_{16}$" from the end of the ram, centre through, drill No. 13 and ream $^3/_{16}$". For the other end I set it up with the back of the ram towards the chuck and was able to drill and ream this hole accurately on the centre height also. Two 2BA Allen grub-screws are fitted vertically to hold the tools.

The index plate is 2" dia. and works in the $^1/_4$" grooves in the ram, so the hole for the spindle must be drilled and reamed $^7/_{16}$" at $^7/_8$" centre from the back face of the ram. I have made the centre height of this hole $^9/_{16}$" from the bottom face of the body to give a little more metal above the hole to prevent distortion when clamped down. This hole must be accurately marked off each end of

the body as it is interchangeable end for end. It is centred each end with a large centre drill and then the $^{27}/_{64}$" drill is held in the chuck of the lathe and, with the back centre in the tailstock, the hole is drilled halfway from each end and then reamed $^7/_{16}$".

At $^3/_8$" from each end of the body and on this $^9/_{16}$" centre height, two $^3/_{16}$" holes are drilled and reamed for the index pins. These pins are made from $^5/_{16}$" silver steel and are hardened right out. The end is made spherical and the head at $^1/_{16}$" thick is slightly domed. A shallow groove the same width as a piece of clock spring is milled between these two holes and a piece of clock spring is bent up similar to the drawing to exert some pressure on these two index pins. Two retainer plates are fitted as shown, made from $^7/_{16}$" b.m.s. with 4BA cap screws, which clamp the clock spring tightly in the groove.

The index plate from 2" b.m.s. is $^1/_4$" thick on the flange with a $^3/_4$" dia. boss $^1/_2$" long and reamed $^3/_8$" for the spindle. Three holes at $^5/_8$" centre distance are indexed with the dividing head and drilled and tapped $^1/_4$" × 32 t.p.i., and a $^1/_4$" Allen grub-screw is fitted through the boss opposite one of these screw holes; this position is important. The index shaft to fit this index plate is $^7/_{16}$" dia. to fit the reamed hole in the body, and is 2" long and $^{11}/_{16}$" long at $^3/_8$" dia. to fit the plate and there is a $^1/_4$" BSF thread $^7/_{16}$" long plus a washer and retaining nut.

On the length of the shaft are three grooves milled for indexing the three positions. These three grooves are indexed accurately and are milled with any suitable V-shaped grooving milling cutter. I used a small involute cutter which has nicely rounded edges and to get the best depth a $^3/_{16}$" steel ball was laid in one of the grooves and a measurement taken over the ball and the shaft. If this measurement approximates 0.585"–0.590" it will be found to work alright. The shaft, after making an indentation for the grub-screw, is case-hardened all over and the thread blued. The indentation for the grub-screw is made in one of the grooves, which are carried right along.

The three adjusting screws are made from $^3/_8$" b.m.s. and are threaded $^1/_4$" × 32 t.p.i. so that each turn represents $^1/_{32}$" of adjustment. The heads of the screws are made $^1/_4$" wide and one screw has three grooves turned in the head, the other with two grooves and the last with one groove. This is so that they are easily identified when graduating. The thread part is $1^1/_{16}$" long and they are case-hardened on the ends of the threads, the ends being reduced in size to $^3/_{16}$". A fourth screw is made, but this is drilled through $^1/_8$" and is to act as a drilling jig for drilling in the correct place for the two fixed stops. Three hexagon nuts are made for the adjusting screws from $^5/_{16}$" A/F hexagon steel and they are $^3/_{16}$" high. The two fixed stops are made from $^1/_4$" A/F hexagon steel, threaded 2BA $^3/_8$" long and stand $^1/_2$" long to the top which is turned to $^3/_{16}$" dia.; the top is slightly domed. These are also case-hardened on the ends.

With the index plate in place and the index pins in one of the grooves, the extra screw is put through the index plate with one of the locknuts to hold it securely.

An $^1/_8$" drill is put through to mark the position on each end of the body and this position is then drilled and tapped 2BA for the fixed stop. The spindle retainer is made of $^3/_{32}$" ground stock to fit the turned groove in the spindle and the milled slot in the body and is made just sufficiently long to clear the spindle when withdrawn and slotted for the $^1/_4$" BSF screw just sufficiently when in the forward position so that it does not bind the spindle.

It remains now to grind the two pieces of H.S. tool steel $^3/_{16}$" dia. to 30 deg. included angle with a very small honed flat. The accuracy of the grinding can be tested on a scrap piece of steel held in the chuck by making a mark and then reversing the tool, when the tool should follow in the same mark.

The fixture can be set up in seconds just like an ordinary lathe tool, the topslide does not need to be set to the angle. The index plate is slipped into the right-hand end when graduating to the left and outwards from the centre, and in the left-hand end when graduating to the right or towards the centre. It will be found to be quick and accurate.

CHAPTER 13

A Quick-Change Toolholder

Every type of toolholder fitted to the centre lathe has disadvantages, some of them serious. An attempt is generally made to rectify some of these faults with special tools, such as the Myford $1/2$" square shank tool with the boat holder. This, once it is set to the correct height, is excellent, but changing the tool and resetting every time to the proper height which generally requires two or three attempts is not quite so good. The other type using a $1/4$" round tool cutting more or less on the end is not much better and is too small for serious turning, and it is even more difficult to change the tool. The four-tool turret using $3/16$" square tool bits is quite good for small work, but again cannot cope with heavy work, also one or other of the tools generally gets in the way and when using a small boring tool, three tools only can be used; the boring tool in any case is quite inadequate for most work.

The range of boring tools are also difficult to set to correct height and in any case, whenever a tool is changed, all readings on the cross-slide and topslide are lost. The operator is always reluctant to change a tool and is inclined to make do with one tool if possible, even though it is not always the most suitable.

I have given this matter a lot of thought and, while there is no easy way out, I have designed and made a quick-change tool for the toolpost which comprises one mounting block and thirteen special toolholders, these can be added to if the need arises. Any tool can be changed in a matter of seconds, whether it is $1/2$" sq., $5/16$" sq., screw cutting or acme cutting, boring from $1/8$" dia. to as large as you like, swivelling tools, and such things as knurling, dial indicator, etc. are available immediately without setting.

As the tools are pre-set, one merely drops the toolholder into place, clamps with the lever, and commences to turn immediately. If a reading has been taken previously with this tool, the same reading is again available. So far I have not found a single disadvantage with this tool, in fact, changing a tool now becomes a pleasure, and used in conjunction with the tailstock, quick-change tools and the saddle feed stop, one has a really excellent turret lathe without a lot of tools sticking out and getting in the way. I may possibly carry this idea a stage further and design roller box tools to fit the mounting block which can use end cutting tools and turn to the full length of the saddle feed. However this may be for the future.

To be successful, accuracy in machining both the mounting block and toolholder is essential. To achieve this accuracy, a properly made drilling jig and special depth gauge are required. These will be described first. The materials required are:

Mounting block: $2^{1}/_{2}$" (finished length) of 2" × 1" b.m.s. Same size in $" plate. $2^{1}/_{2}$" of 1" key steel for the tongue.

I have named the toolholders type A, B, etc., for easy reference.

Type A. – 5 off: 14" of $1^{1}/_{4}$" key steel or $1^{1}/_{4}$" × 1" b.m.s. ($1/2$" sq. tools).

Type B. – 2 off: 6" of $1^{1}/_{4}$" key steel ($5/16$" sq. tools or odds).

A QUICK-CHANGE TOOLHOLDER

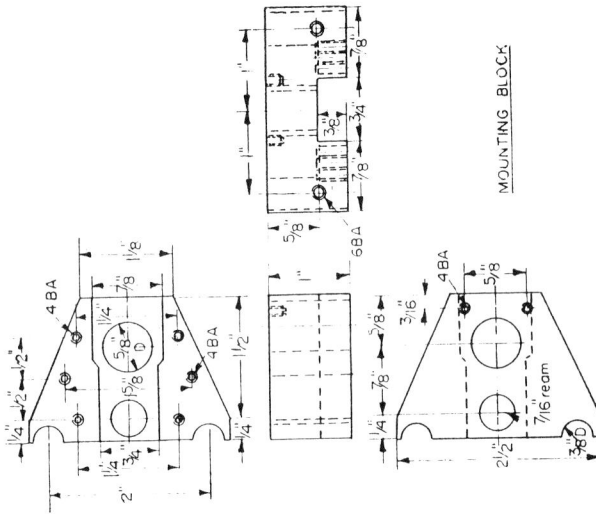

MOUNTING BLOCK

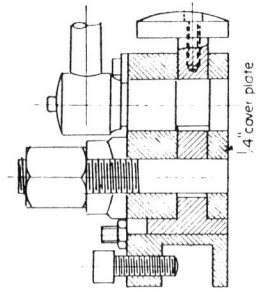

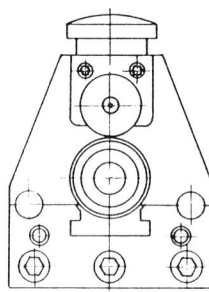

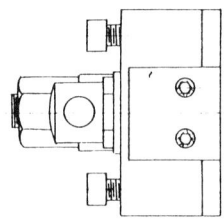

ASSEMBLY OF TOOL HOLDER AND MOUNTING BLOCK

Type C. – 1 off: 4" of $1^1/_4$" key steel (swivelling $^5/_{16}$" sq. tools).

Type D. – 2 off: $8^1/_4$" of 1" key steel (screw cutting tools).

Type E. – 2 off: $7^1/_4$" of $1^1/_4$" key steel ($^7/_{16}$" and $^3/_8$" boring tools).

Type F. – 1 off: 3" of $1^1/_2$" key steel or $1^1/_2$" × $1^1/_4$" b.m.s. ($^9/_{16}$" boring tool).

In addition materials for drilling jig, silver steel for boring tools, Allen screws, etc. as will be described.

The drilling jig is made of a $4^1/_4$" piece of 2" × $^1/_2$" b.m.s. and is drilled mounted on a milling table in the drilling machine, in the vertical mill or on a vertical slide in the lathe. A wobbler or other means of locating the edge of the plate is necessary as ordinary marking out is not sufficiently accurate. At exactly 1" from one edge and at 2" centres in the position as shown, allowing for the stop block of $^1/_2$" key steel, the two letter U holes are drilled, centre drill first, then small drill, then the letter U drill, completing each hole with the slides locked before moving to the other one. These holes are for the locating pegs and are most important. The next two holes towards the edge are at $1^3/_4$" centres and are drilled No. 24 and are for the tapping size of the height adjusters. The next three holes are the clamping screw holes and are drilled at $^{15}/_{16}$" centres No. 4, for $^1/_4$" Whit. tapping. Right in the centre are two $^5/_{16}$" holes at $^7/_{16}$" centres and $^{13}/_{16}$" from the same edge, these are to remove metal for end milling for the clamping tongue and the edges of these holes are used to locate for the end mill, so must also be accurate. Take all the measurements using the machine slide graduations, not only from the one edge, but also from one end, in order to obtain accuracy in the middle. The stop block on one end is secured with a couple of 2BA Allen cap screws. The letter 13 drill is the reaming size for the $^3/_8$" silver steel locating pegs. The drilling jig is not reamed of course, but all drill sizes must be as stated. This drilling jig is now case hardened thoroughly.

Drilling the toolholders

Type A toolholder: If made from $1^1/_4$" key steel, a short piece of $^3/_4$" key steel will be required alongside the $1^1/_4$", and the two with the drilling jig on top clamped in the machine vice, the fixed jaw against the drill jig and the $1^1/_4$" key steel and a piece of wood to clamp up against the other jaw. If no large machine vice is available, two pieces of 2" angle iron with a couple of $^3/_8$" bolts will do almost as well, but clamp as in the machine vice so that the locating edge of the jig and the $1^1/_4$" key steel are flush. Drill all the holes, two letter U, two $^5/_{16}$", two No. 24 and three No. 4. If the Type A is made of 1" × $1^1/_4$" b.m.s., two toolholders can be clamped together under the drill jig; make sure the edge is dead central with the two letter U. You will note that the No. 4 drills for $^1/_4$" Whit. tapping are at $^5/_{16}$" from the edge, this is for added clamping strength rather than being located in the centre of the $^5/_8$" milled tool slot.

The two Type B tool holders are drilled edge to edge in the centre and here two pieces of $^1/_4$" key steel will be required to locate the two pieces of $1^1/_4$" key steel in the centre. Use a piece of wood alongside the undrilled half of the drill jig, which should locate the pieces correctly. All holes are drilled as in Type A except that the three No. 4 holes are this time drilled at $^1/_4$" from the edge. This is done by locating a piece of $^1/_{16}$" thick strip alongside the drill jig.

Type C, one off, is drilled as Type B, but no No. 4 holes are drilled. Type D, two off, are made from 1" key steel and are drilled both together under the drill jig, but again no No. 4 holes are drilled. Type E, two off, are also drilled two together, locating one against the stop block, but allowing the other to pass an equal amount past the block. In the Type D, the stop is removed and both pieces are centrally located. The type F, one off, should first be milled or turned down to $1^1/_4$" thickness if made from $1^1/_2$" key steel and is drilled centrally with stop block removed, no No. 4 holes are required. I have drilled my boring

A QUICK-CHANGE TOOLHOLDER

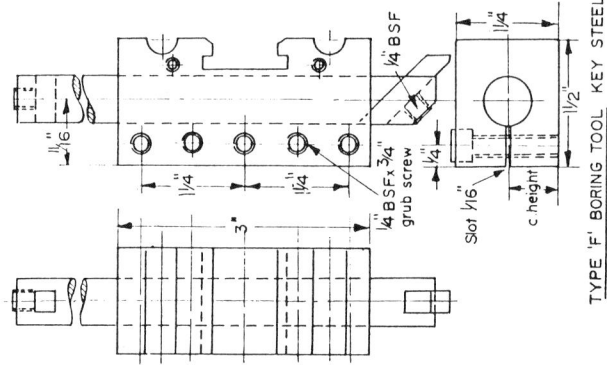

TYPE 'F' BORING TOOL KEY STEEL

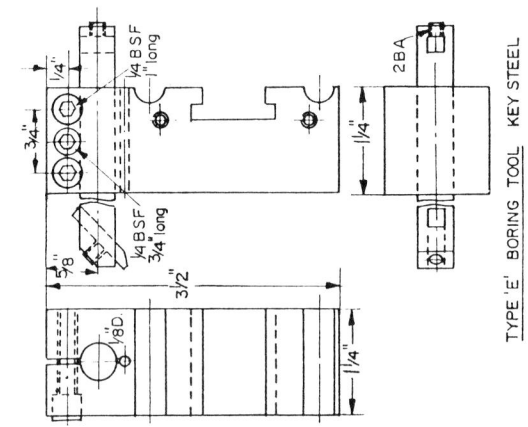

TYPE 'E' BORING TOOL KEY STEEL

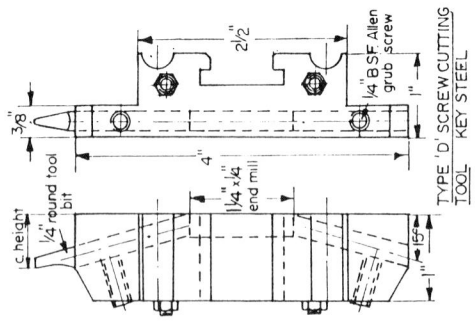

TYPE 'D' SCREW CUTTING TOOL KEY STEEL

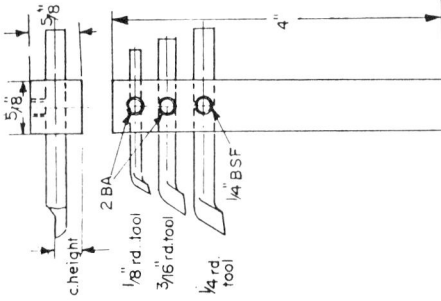

SMALL BORING TOOL HOLDER FOR USE WITH TYPE 'B' TOOL HOLDER

5/8"
5/8"
c.height
2 BA
1/8 rd tool
3/16 rd.tool
1/4 rd. tool
1/4 BSF
4"

1 off L = 1"
1 off L = 9/16"
3/4 D
1/4 x 40 T
25 divisions
Araldite
L
5/16 D S.S.
2"
5/8"
2 1/2"
1"
DEPTH GAUGE

2 5/8"
c height
3 5/16"
1"
1/4"

7/10 A.F
c.height
1" D x 1/8" washer
1/2"
1/4"

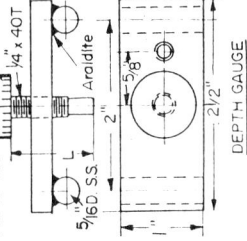

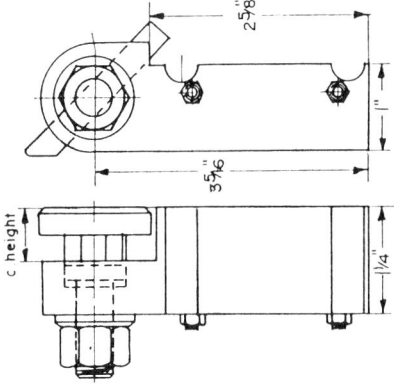

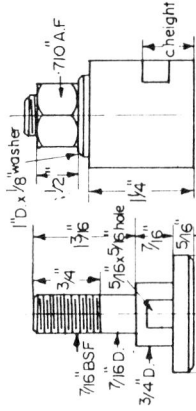

TYPE 'C' SWIVELLING TOOL HOLDER FOR 5/16" SQ. TOOLS
KEY STEEL

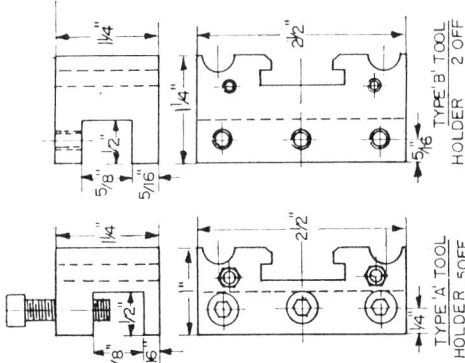

CLAMP BOLT B.M.S
7/16 BSF
7/16 D
3/4 D
3/4"
5/16 x 9/16 hole
1 3/8 D
13/16"
7/16"
5/16"

1 1/4"
1/4"
2 1/2"
5/8"
5/16"
1/2"
5/16"
TYPE 'B' TOOL HOLDER   2 OFF

1 1/4"
1"
2 1/2"
5/8"
3/16"
1/2"
1/4"
TYPE 'A' TOOL HOLDER 5 OFF

125

toolholders with the height adjuster holes, but these are not really required as these toolholders are always set with the bores central with centre height, as will be explained later.

Now drill the No. 24 holes *from the bottom* to $^{3}/_{16}$" to within $^{3}/_{8}$" of the top edge and this remaining portion can be tapped 2BA. It is important that these holes be tapped *from the bottom* through the $^{3}/_{16}$" size, the tap will go through alright, as it will be impossible otherwise to get the thread perfectly square with the rest of the hole. One thing I have forgotten to mention in drilling all these holes, I did not drill any of them right through while held in the vice, to avoid raising burrs, but set stops and drilled to within $^{1}/_{16}$" of the bottom face. The holes were then finished over one of the slots in the drilling table, removing burrs each time as necessary. The letter U holes are now reamed $^{3}/_{8}$" and, in the case where two pieces were clamped together, the drill shank was clamped up in one pair of holes while reaming was done in the other. Then a piece of $^{3}/_{8}$" silver steel was clamped in the reamed hole while the other was reamed. A machine reamer must be used, a hand reamer is useless for this work.

A $^{3}/_{8}$" hole was drilled longitudinally through where the slot for the tool will go in the Type A and B toolholders; this was done to make the milling of these $^{5}/_{8}$" wide by $^{1}/_{2}$" deep end milled grooves easier. Before that huge pile of drilling swarf is cleared away, the mounting block should be drilled using the same drill jig. As the two locating pegs are drilled $^{1}/_{4}$" from one of the two longer sides and are drilled with the bottom $^{1}/_{4}$" steel plate in place, this plate must be fixed in position and all machining on the outside of the mounting block done with this in place. Mark out the shape of the mounting block on the $^{1}/_{4}$" plate and drill the six No. 32 holes for the 4BA cap screws which secure the plate to the mounting block. These are at $^{1}/_{2}$" centres parallel with the 2" length and at $1^{1}/_{4}$" and $1^{1}/_{2}$" centres parallel with the $2^{1}/_{2}$" length as shown on the drawing. Clamp the

plate to the block and drill through $^{3}/_{8}$" deep, tap the block 4BA and drill the plate and counterdrill to fit the head of the screws just slightly below flush. Now screw the plate to the block and file the plate roughly flush with the block all over, to make it easy to clamp for drilling.

The mounting block and plate can now be clamped under the drilling jig for drilling the two letter U holes for the locating pegs. A piece of $^{3}/_{4}$" key steel will bring this to the correct location and the stop block is used. After drilling, these holes are also machine-reamed $^{3}/_{8}$". Do not machine the two sides to the angle shown until the very last, as the parallel sides are so much easier to hold for the rest of the machining. The slot for the clamping tongue should now be end milled. Using a $^{5}/_{8}$" end mill, mill right through dead in the centre to $^{3}/_{8}$" deep in the mounting block with the bottom plate removed of course; this is milled across the 2" width and must be central. The width of the slot is $^{3}/_{4}$" and is then enlarged to $^{7}/_{8}$", still using the $^{5}/_{8}$" end mill, for a length of $1^{1}/_{8}$" leaving the balance at $^{7}/_{8}$" long to $^{3}/_{4}$" width at the locating peg end.

The clamping tongue is next made, as this is drilled and reamed in place in the mounting block. The tongue is made from 1" key steel and is milled from the solid to $2^{3}/_{8}$" total length. The head is made a free fit in a $^{3}/_{16}$" milled slot and is a bare $^{1}/_{4}$" below the central part. This central part is a free fit in the mounting block and should be minus 0.010" in width and minus 0.005" thickness; the head is left 1" square and this will stand $^{3}/_{8}$" above the central part.

The tongue should now be fitted in place in the mounting block with a couple of bits of shim steel on the sides and the bottom plate secured in place. The whole should be mounted in the four-jaw chuck, one jaw held firmly against the tongue and the opposite jaw clear of the other end of the tongue by means of packing. One point here that I had forgotten: the face against which the tongue will press should be machined 0.010" off to give clearance, so that the tongue will clamp

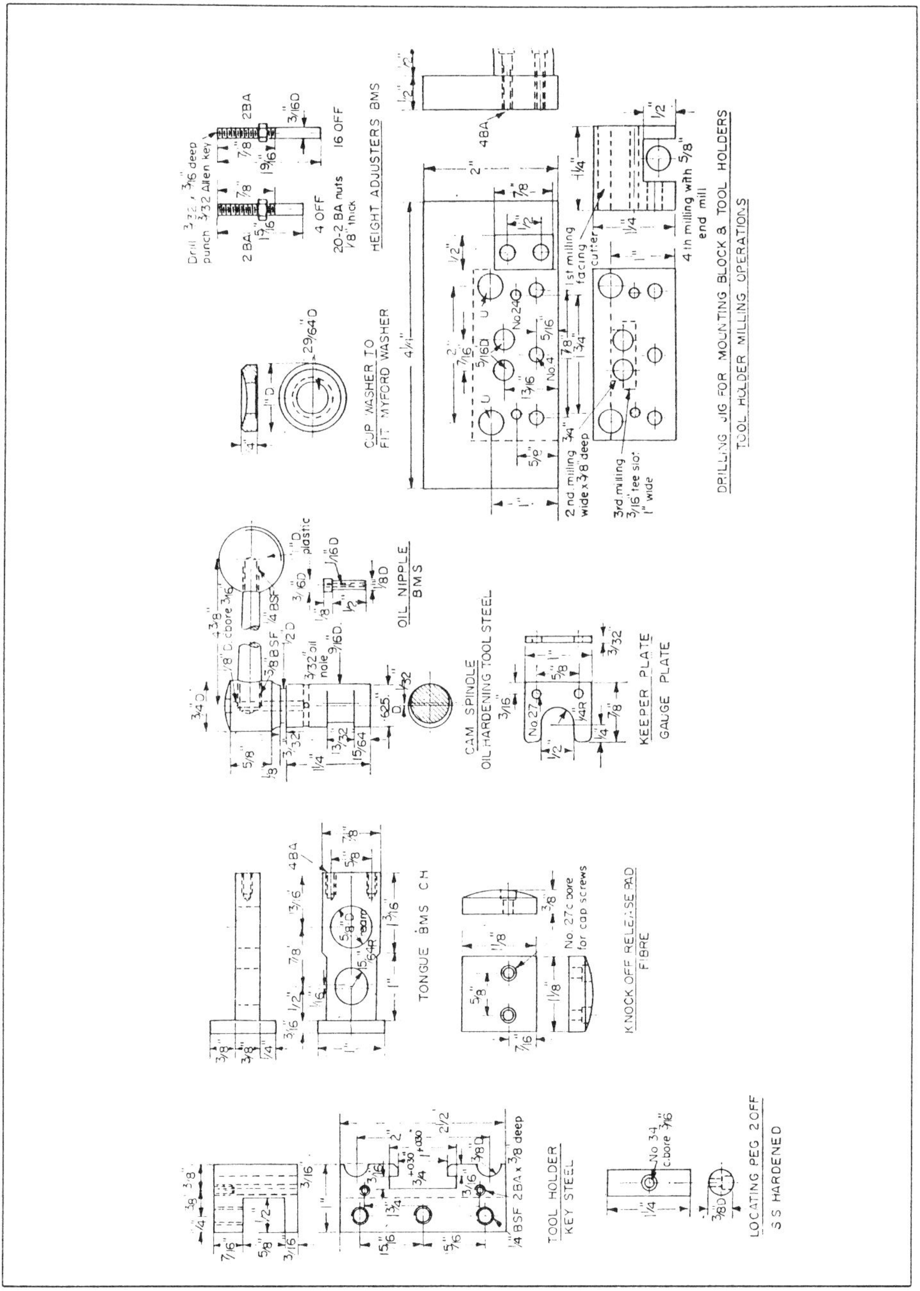
HEIGHT ADJUSTERS BMS
4 OFF
16 OFF
20-2 BA nuts 1/8" thick
CUP WASHER TO FIT MYFORD WASHER
DRILLING JIG FOR MOUNTING BLOCK & TOOL HOLDERS
TOOL HOLDER MILLING OPERATIONS
1st milling facing cutter
2nd milling 3/4" wide x 3/8" deep
3rd milling 3/16" tee slot 1" wide
4th milling with 5/8" end mill
OIL NIPPLE BMS
CAM SPINDLE OIL HARDENING TOOL STEEL
KEEPER PLATE GAUGE PLATE
TONGUE BMS C.H.
KNOCK OFF RELEASE PAD FIBRE
No. 27 c bore for cap screws
TOOL HOLDER KEY STEEL
1/4 BSF 2BA x 3/8 deep
LOCATING PEG 2 OFF S S HARDENED

How the toolholder looks on the lathe.

against the toolholders and not against the mounting block. It is mounted central, so that the centre drill will touch a point $1/4$" from the mounting face, and it is then centre drilled, drilled with a small drill about $1/8$" then drilled $27/64$", lightly bored a few thou and machine reamed $7/16$" for the Myford clamping bolt on the toolpost. Then, moving it along in the chuck (retaining the central position) $7/8$" for the cam spindle, this is centre drilled, drilled $1/8$", drilled $9/16$", bored to 0.620" and machine reamed to $5/8$", or failing a reamer of this size, carefully bored to 0.625" using a piece of silver steel as a gauge. The tongue, meanwhile, is still firmly held against the face with one of the chuck jaws as before.

The tongue is now removed and set up in the four-jaw chuck on its own with the $7/16$" reamed hole central. This hole is now bored to $29/64$" and then moved along $1/16$" in the chuck and the hole elongated towards the other $5/8$" hole. This is to give clearance to the tongue when clamping pressure is removed in order to change the toolholders.

Cam spindle

As it is necessary to complete the mounting block before milling the toolholders, the cam spindle should now be made. This is made from a piece of $7/8$" or 1"

dia. oil-hardening non-shrink tool steel of good quality as used for press tool work. The piece is held in the four-jaw chuck with $1 5/8$" protruding from the jaws and is turned to a close slide fit in the $5/8$" reamed hole in the mounting block for a length of $1 15/32$". The $3/32$" wide groove for the keeper plate is turned to $1/2$" dia. at a bare $1 1/4$" from the end, then the piece is offset in the chuck $1/32$"; using cross-slide graduations to ensure this, a difference of $1/16$" in the readings should be obtained. The end is now centred in this position and the cam turned to $9/16$" dia., not quite cleaning up the high side, commencing $15/64$" from the end and making the cam $13/32$" wide; this is to give end clearance in the tongue. The cam spindle can now be reversed in the three-jaw chuck and the end turned to $3/4$" as in the drawing, the end domed and a $1/8$" oil hole drilled $1 1/4$" deep, lightly counterboring to accept the $3/16$" dia. oil nipple.

On the high side of the cam, the hole for the handle is drilled and tapped $3/8$" BSF, $9/16$" to the point of the drill, and the cross oil hole is drilled $3/32$" about $3/16$" down from the retainer groove. The handle is tapered from $3/8$" b.m.s. to $1/4$" BSF for the 1" plastic knob and the centre-to-centre distance is $4 3/8$". The cam spindle should be hardened in oil and tempered very lightly from the top, leaving the working faces almost dead hard and only the top blued. Now screw in the handle tightly and make and fit the little oil nipple from $^3/_{16}$" b.m.s. with a $^1/_{16}$" oil hole, this will also act as a key for the handle, the oil gun being used for lubrication.

Locating pegs

These are made from $^3/_8$" silver steel and are made exactly to the $1^1/_4$" length of the mounting block and base. They are pushed or pressed into place in the holes and marked 1 and 2, then held if necessary by a clamp on their ends and drilled in the centre of the two 6BA Allen cap screws. The holes are drilled tapping size first, right into the mounting block about $^3/_8$" and also the clearance size through the pegs and finally counterbored for the heads of the 6BA cap screws, all without removing from the mounting block. The holes can also be tapped before removing the pegs from the mounting block. They are now removed and hardened in oil and tempered to a just perceptible trace of colour.

The mounting block, complete with bottom plate, is now milled on the mounting face 0.250" off, leaving the central part where the tongue fits at 0.750" wide; this must be quite central as some of the toolholders are reversible for left and righthand cutting. As the face was previously machined 0.010" off, this new face will also be 0.010" below the central line, which will give the necessary clearance to the toolholders. Before leaving the mounting block, make the retaining plate to the drawing from $^3/_{32}$" gauge plate and secure it in place so that the cam spindle is perfectly free, with two 4BA cap screws, and also mill the two sloping sides leaving the end at $1^1/_8$" wide and tapering out at $^1/_4$" from the clamping face.

If you are confident that everything is now perfect, the whole mounting block can be case hardened, paying particular attention to the $^5/_8$" reamed hole in both the mounting block and the bottom plate. Then lap the bottom face of the mounting block to

perfect flatness on a cast-iron plate and also both sides of the bottom plate. The outer end of the tongue piece is drilled and tapped for the two 4BA cap screws which hold the fibre knock-off release pad. This is necessary as the toolholders clamp tightly to the two locating pegs only, the faces being 0.010" clear, and even with the lightest pressure, grip very firmly and after releasing the cam lever, require a bump with the heel of the hand to free them. If the head of the tongue piece is quite free in a slot milled with your $^3/_{16}$" T-slot cutter, it can have 0.005" clearance. The tongue piece can also be case hardened, again especially around the $^5/_8$" reamed hole and the head.

The toolholders

Now comes the rather long job of milling all the toolholders to fit the mounting block. The mounting block itself is used as one jig to make sure the central slot is truly central, so that the mounting block will clear this slot when tried either way. The edges of the two $^5/_{16}$" holes already drilled will give a very good indication of the centre, besides leaving less metal to remove, but the depth of the slot is very important, as all the toolholders must clamp tightly with the cam lever in the same position, that is at about 30 deg. from parallel with the lathe centres. A special depth gauge is required here and a piece of 1" × $^1/_4$" b.m.s. with two $^1/_4$" × 40 t.p.i. tapped holes for the gauge screws – as shown on the drawing – should be made. The two locating pegs are not $^3/_8$", but are two pieces of $^5/_{16}$" silver steel as these will bottom in the $^3/_8$" reamed holes of the toolholders with no trouble. These two pegs are secured to the face of the steel by means of Araldite and are located by resting the whole piece in one of the toolholders overnight with a weight to hold the silver steel pegs square and at the correct centres. The two screws are screw-cut and threaded with dies to a close fit in the tapped holes and the short one is for measuring the face of the toolholders from the bottom of the peg slots. The longer one is for measuring the bottom of the recess to the bottom of the peg slots. The heads are graduated with 25 divisions and a zero mark is made in the centre of the steel plate.

All those toolholders which require machining down on the clamping face by $1/4$" are machined first. I use the $5/8$" end mill run at 500 r.p.m., flooding with soluble oil and water. If done in the lathe, a speed of 425 r.p.m. with the pressure of the feed downwards will prove effective. The gauge can be set to one of those toolholders which are already down to size, or better still, one holder can be completely machined, faced, grooved down and T-slot grooves cut and tried in place on the mounting block and rectified where necessary; then the gauge can be set to this toolholder to bring all the rest to the same setting. This was the way I did it, and it proved completely satisfactory.

For the bottom of the grooves, I slotted this 0.005" more than necessary so that the T-slot cutter did not have to bottom on this while cutting the T-slots. After the facing is done, the recess is milled to the gauge and to 0.020" clearance on the sides of the tongue piece left in the centre of the mounting block and then, without shifting each toolholder, it is better to mill the T-slots by changing over to a T-slot cutter. I used one I had made for T-slots $3/16$" wide and which would cut a full $1/8$" deep before rubbing on the shank. This was set to a 0.005" feeler gauge to the bottom of the slot and so, provided the slot was the correct depth to the gauge, the clamping position of the T-slot was also correct.

After all the toolholders are milled to fit the mounting block, the type A toolholders can be finished off. I milled the first one $5/8$" wide again (using the $5/8$" end mill) and $1/2$" deep, leaving the bottom thickness exactly $3/16$". I was a little diffident as to whether this thickness was enough, but you will notice that the clamping screws are close to the inside face of the slot. After tapping for the 1" × $1/4$" Whit. Allen cap screws, I measured the depth of the toolholder with the micrometer and then, after clamping a tool

sufficiently tightly in the holder, I measured it again. The expansion was only about one or two thou, which I considered did not matter, so I went ahead with the whole five Type A holders. Then, after dressing the bottoms of my set of Myford $1/2$" shank boat-holder type tools to make sure they were truly flat, I set one up in the lathe, expecting to have to grind a little off the top of the tool to bring it to centre height. To my surprise and delight I found that every tool came exactly to the centre height so I was really lucky.

For the Type B toolholders, two off (one of which I made 3" long and put four clamping screws in it), I measured the thickness of the bottom of my four post turret and found it to be 0.350". After checking that the $5/16$" square tools did come truly to centre height, I made these two toolholders to this thickness at the bottom of the tool slot, again using the $5/8$" end mill to $1/2$" depth. In this case, the clamping screws are $1/4$" from the end and so come to the middle of the slot, allowing the tool to be swung at an angle if required.

One of these Type B toolholders, the short one, is used for general purpose tools, such as the toolholder made from $5/8$" key steel for holding three small boring tools $1/8$", $3/16$" and $1/4$" as shown on the drawing, the knurling tool and the shank of the dial indicator, etc. The other one, being longer, allows the small tools to get closer to the centre on small work, these tools being the set of tools normally used in the four-post turret. If the need arises, I may make one or two more of this type B and indeed, a few more type A would be an advantage as with only five of these, this allows three tools right-hand for steel, one left-hand for steel and one tungsten carbide for cast-iron and non-ferrous metals; not by any means too many!

Another holder for $5/16$" tools, which also holds $5/16$" shank boring tools, is the swivelling toolholder Type C. This is a very useful tool for all those awkward jobs which require putting the tool at awkward angles such as undercutting, etc. In this case, the

Three boring tools and the swivelling toolholder.

face against which the top of the tools clamp is 0.350" plus $^5/_{16}$" or 0.6625" from the bottom. This cut-out can be milled and the rest finished in the drilling machine as I did, or can be machined and drilled and counterbored in the four-jaw chuck. The hole is $^7/_{16}$" for the clamp bolt but is counterbored $^3/_4$" dia. to $^3/_{16}$" deep to clear the clamp screw, the metal above the $^5/_{16}$" sq. hole being only $^1/_8$" thick, the bottom of the square hole being slightly chiselled out below the face of the head, so that clamping is between the two outer faces. The thread of the bolt is $^7/_{16}$" BSF.

The two Type D screwcutting tools replace the two tools I made from 1" × $^3/_8$" steel. The total length is 4" and the tools are $^1/_4$" round H.S. steel. The holes are drilled downwards at an angle of 15 deg. and, where they meet at the bottom, a slot is end milled $^1/_4$" wide and $^1/_4$" deep by 1$^1/_4$" long, to make it easy to remove the tools. The reason that these tools are made double ended is that it is often better to have the tool close to the tailstock when screwcutting a thread on the end of a shaft, and so the whole mounting block is swung round to face the tailstock and the tool bit is placed in the other end of the holder. The tool bits *do not* have a flat ground on them and the $^1/_4$" Allen grub-screw has its tip flattened off so that the tool can be set to the helix angle of the thread being cut, whether right or left

handed. The reason that two toolholders are made is so that one tool can be ground to 55 deg. and the other to 14$^1/_2$ deg. for Whitworth and Acme threads. The tools are ground parallel on the top leaving $^3/_{16}$" depth; this will then leave a top rake of 15 deg. and the position of the hole must bring the tool to the centre height. This makes a very free cutting tool and, being thinned down to $^3/_8$" for a length of $^3/_4$", allows the tool to be used close to a shoulder.

The reason that the cam lever is placed in line with the high point of the cam on the cam spindle is so that it can be used to clamp up when moved to the right normally, but it also can clamp up when moved to the left, when the mounting block is reversed on the toolpost, or turned around parallel to the lathe bed, as when using the heavy boring tool.

The two type E boring tools are for a $^7/_{16}$" boring bar 5" long and for a $^3/_8$" boring bar 4$^1/_2$" long. The $^7/_{16}$" one takes $^3/_{16}$" sq. tools, one at 45 deg. and the other square across, and the $^3/_8$" bar takes $^1/_8$" sq. tools. 2BA grub-screws are used in each case. Two 1" × $^1/_4$" cap screws clamp the bar and one $^3/_4$" × $^1/_4$" grub-screw in the centre forces the hole open to make it easy to move the bar. The cap screws are brought to within $^1/_{32}$" of the edge of the boring bar hole in each case, as the toolholder is of heavy section and difficult to clamp. The

131

centres of the cap screws are $1/4$" from the end and sides of the toolholder. The $7/16$" and $3/8$" holes were drilled and reamed right on the centre line while held in place on the mounting block. I considered whether to raise the hole above the centre line to make provision for the thickness of the tools, but decided to grind a top rake on the tools and then twist them down to the centre line, which means that the boring bars can enter a hole very little larger than the bar itself.

You will have noticed that there is a concave washer shown on the drawing to fit the Myford clamp washer, this is 1" dia. and $1/4$" thick and of course the Myford clamp nut is used. It will be necessary to fit these on the mounting block before the boring tools are made and indeed the Type A and B can be end milled in place for the tools, provided that the whole mounting block is packed up to the correct height.

The Type E tools are slotted $1/16$" into the bar holes and I also drilled a $1/8$" hole as close in as I could and carried the slot into these holes. I had to use the hacksaw for this last bit as I did not have a large enough slotting saw. It is very necessary to get a very close fit around the bars and I spread the hole to the maximum I could and then lapped in the bars. After cleaning up, a slight pressure on the grub-screw made the bars free to slide and they were easily clamped up tight. The bars were my Eclipse boring tools, but if these are made, it is better to use silver steel for these rather than ordinary b.m.s.

The Type F boring tool, one off, is 2" wide to allow sufficient metal for the $9/16$" dia. bar which is 7" long with two $1/4$" sq. tools,

the clamping screws being $1/4$" Allen grub-screws. Here again the 1" cap screws are brought to within $1/32$" of the bar hole and there are three cap screws 1" × $1/4$" Whit. and two $3/4$" × $1/4$" grub-screws for spreading. After boring on the centre height I finished with a $9/16$" drill as I did not have a reamer of this size, and then spread the hole as much as possible with the two grub-screws screwed very tightly; this bar was also lapped in. The result is excellent, the bar being free with grub-screws tightened and clamps very securely quite easily. This tool was also ground with top rake and twisted around to the centre height and so will enter a $5/8$" hole, whereas if the tool were $1/8$" below the centre height to allow for the thickness of the tool, it would be difficult to enter the tool in anything under a $7/8$" dia. hole. The heads of the $1/4$" Allen screws in these boring bars are brought flush with the toolholder for neatness.

As all the tools are pre-adjusted to the correct centre height, there is no need for the height adjusters until the tools are ground down to below the centre height. These, however, were made and fitted, and the bottoms brought just clear of the bottom faces of the toolholders. They are made from $3/16$" b.m.s. and the threads are 2BA $3/4$" long. They must of course be threaded in the dieholder in the lathe so that the threads run true. Sixteen will be required, finished length $1 9/16$" and four of a finished length of $1 5/16$"; these were all made $1/16$" longer and the end first centred and drilled $1/16$" deep No. 37 and then a further $3/16$" deep to $3/32$". These holes are punched out to a hexagon hole to fit the $3/32$" A/F Allen key, and the No.

Eight type A toolholders.

37 hole is to guide the punch truly into the hole. The punch is made from a short length of Allen key, cut off and ground flat and sharp on the end. The pieces of $3/16$" rod are all cut off to length unscrewed, and are punched before trimming off the end and threading. To hold the pieces for punching, I gripped two short ends of $3/4$" key steel together in the machine vice with a piece of card between them and drilled through $3/16$". This gripped the pieces firmly and it was but a matter of moments to punch the pieces. They were then put back in the collet chuck in the lathe and the end reduced nearly $1/16$" and threaded $3/4$" to a free fit in the toolholders. These and the nuts proved to be a nice little job for the tailstock quick change toolholder. The nuts are made from $1/4$" A/F hexagon steel and are $1/8$" thick and tapped.

Lathe tools are perhaps the most important part of a lathe and yet they seem to have received very little consideration by lathe designers. It was only after using this Quick-Change Toolholder for a few days that I realised how much time and patience I must have wasted over the years using archaic and crude methods of holding lathe tools. One could now carry this a stage further and, by dispensing with the topslide for repetition work only, and mounting the quick-change holder on a specially designed table directly bolted to the cross-slide, with its own built-in indexing to two or three positions, and with roller box and other turret tools, the lathe would be capable of a lot of quick work. With this method there would be sufficient depth below the centre height for end-cutting tools in the roller box tool holders, which could be used with power feed for the length of the saddle feed.

A QUICK-CHANGE TOOLHOLDER

CHAPTER 14

Stop Bars and Bushes for Lathe Mandrels

The stop bars for the lathe mandrel here described are really three different fixtures. The length stop with three lengths of stop bar is held in position by means of a holder at the left-hand end of the lathe mandrel and adjusted quite simply from outside after the first piece of material is inserted and set for the best position; subsequent pieces will then be positioned exactly the same. Pieces the full length of the mandrel can thus be machined to the same lengths and setting is no trouble, being set from the left-hand end.

The second fixture is really a set of bushes to support the end of the material being turned, which takes the strain off the collet or chuck jaws and prevents that flapping around of long lengths being machined in the chuck. It uses the same holder on the left-hand end of the mandrel as the stop bars.

The third fixture is for facing off thin discs on the other side after parting off. A $3/8$" steel rod is supported just behind the chuck jaws in a Morse taper sleeve which is drawn into place in the lathe mandrel by means of a steel tube and a knurled nut on the left-hand end of the mandrel; this nut locates the $3/8$" rod for length and on the chuck end of the rod are fixed a number of turned discs against which the job is held to run true and the thickness required can be maintained without trouble.

The three fixtures were made in a very short time from short offcuts in the scrap box and have repaid the time to make them many times over. The dimensions shown are for the Myford Super 7, but they can easily be changed to suit any lathe. On the

end of the lathe mandrel is a ground journal 0.8685" × $1/2$" long in my case and use was made of this for the bar and bush holder.

This holder is made from a piece of $1/2$" b.m.s. bored to a slide fit $9/16$" long by 0.869" dia. and bored to fit the three stop bars a slide fit. The total length is $1^3/16$". It is knurled on the $1^1/2$" diameter and drilled and tapped for a $1/4$" BSF grub-screw in the centre of the $1^1/4$" diameter part to hold the stop bars in location. This grub-screw also holds the material support bushes. The $1^1/2$" diameter part is slit into the centre hole with a $1/16$" slitting saw and it is also slit through half the diameter at the end of the $1^1/4$" diameter part as shown on the drawing. A 2BA cap screw is drilled at right-angles to the first slit to close up the fixture and clamp it tightly to the end of the lathe mandrel. A $5/16$" end mill is let into the part at $3/8$" from the centre of the saw cut to $5/16$" depth and so drilling for the cap screw becomes simple; the bottom half of the piece is of course tapped.

It would be better to make the three stop bars before the holder is made, so that when the bars are made a free sliding fit in the lathe mandrel, the smaller end of the holder can be made to fit the stop bars. I used $5/8$" b.m.s. for the three bars and reduced it with a lathe file down to 0.621" which was perfectly free in the mandrel except for the end nearest the chuck which was reduced to 0.580" for 4" only on the longest bar only. This was to clear the bore of the mandrel where it is reduced for the end of the Morse taper. The ends of the stop bars are centred to clear any pip that may

STOP BAR B M S   3 OFF

1 off L = 14"
1 off L = 8½"
1 off L = 4"

¼" flat

5/8 D.

1/8" 9/16"
1"D   d'   5/8 D.

1 off each 'd' =
½" 7/16" 3/8" 5/16"
¼" 3/16" 1/8"

BAR SUPPORT BUSH B M S

¼" B S F grub screw
Knurl
½"D
1/4 D
5/8 D.
.869" bore
1/16 slit
2 B A
5/8" 9/16" Slit to halfway

STOP BAR AND BUSH HOLDER   B M S

Drill ¼" & centre drill
3/16" flat
3/8 D.
16"

STOP BAR FOR HOLDING STOP DISCS   B M S

3/4"
½" x 32 T. x 1¼" long
½"D
½"
½" x 32 T.
No. 2 Morse
1½"
7/8"
3/8 bore
10"
3 5/8"
End of mandrel
End of mandrel
Face of chuck

STOP BAR SUPPORT AND DRAW IN TUBE

1"D   Knurl
½" x 32 T
3/8 D
7/8 D
5/8 D
¼" B S F
5/8" 1" 1/8"

NUT FOR DRAW IN TUBE

'd'
't' 3/8"
1/8"
'D'

STOP FOR FACING DISCS

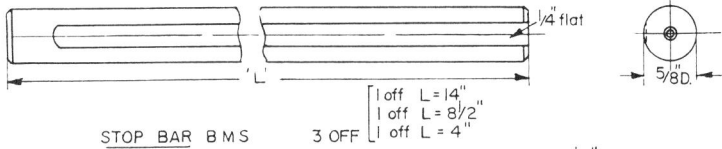

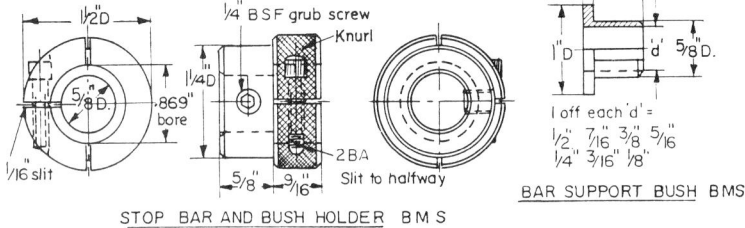

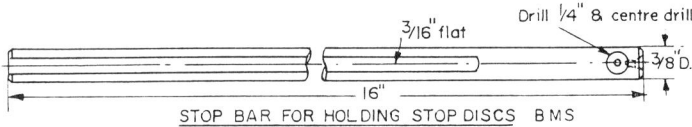

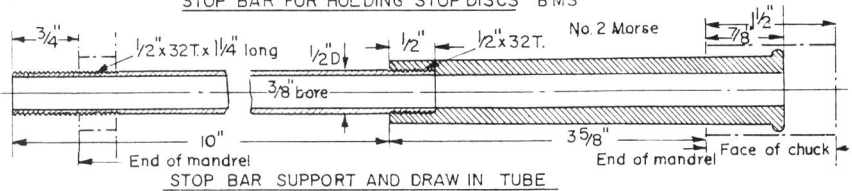

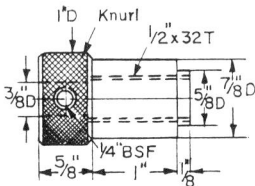

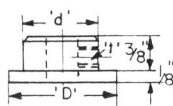

D	d	t	No. off
2"	7/8"	¼ B S F	1
1½"	"	"	1
1¼"	"	"	1
1"	"	"	1
3/4"	23/32	2 B A	1
5/8"	19/32	"	1

be left on the end of material being used. A flat ¼" wide was filed along most of the length of each bar for the grub-screw to bear against, so that burrs are not raised on the diameter. These flats were finished by draw-filing with a fine file.

The bar holder can now be bored to a nice fit for the bars. The bushes are all made from short ends of 1" dia. b.m.s. and are turned a push fit 9/16" long to fit the bar holder. I made the smallest 1/8" and the largest ½", each rising by 1/16". The end which fits into the holder was deeply tapered in the hole so that the material would easily locate and enter the hole. A similar flat was filed on each one. The stop bar for thin discs is a little more elaborate, but is so easy to use and so useful that it is well worth the trouble. Discs which need to be faced off on the side which has been parted off from the bar, can be faced to an accuracy for parallelism and thickness that looks perfect with the micrometer. Make the Morse taper sleeve from $7/8$" b.m.s. and let it protrude from the nose of the lathe so that it just clears the inside of the three-jaw chuck jaws. It is bored and reamed $3/8$" which will make it a free sliding fit on the $3/8$" stop bar. The smaller end is tapped $1/2$" × 32 t.p.i. for the draw tube and the larger end is made with a small ridge to make it easy to hold. The $1/2$" steel tube is a full $3/8$" bore and is screwed $1/2$" long and screwed tightly into the Morse taper sleeve. The other end has $1\frac{1}{4}$" thread and should protrude from the lefthand end of the lathe mandrel by $3/4$".

The knurled nut for the draw tube is made from 1" dia. b.m.s. and is bored and tapped $1/2$" × 32 t.p.i. for a depth of $1\frac{1}{8}$" and the balance is drilled and reamed $3/8$" for the stop bar. The total length is $1\frac{3}{4}$". The $3/8$" hole part, which is the 1" dia. knurled end, is fitted with a $1/4$" BSF grub-screw and is turned down to $7/8$" for 1" and for $1/8$" to $5/8$" to fit into the end of the lathe mandrel to locate it truly.

The stop bar which is located for length in the knurled nut just described is 16" long and has a $3/16$" flat filed along part of the length. The other end which goes through the tapered sleeve has a deep centred hole at $5/16$" from the end to take the grub-screws in the stop discs. The end is also centred to clear any pip, as this stop bar without the discs can be used for short lengths held in the chuck where the pieces are from $3/8$" up to $5/8$" dia.

The stop discs are all drilled and reamed $3/8$" to fit the stop bar and are fitted with an Allen grub-screw. They are turned to an outside diameter of 0.010" to 0.015" smaller than the nominal size as shown on the drawing, so that the chuck will grip on the work and not on the stop discs.

The boss is $7/8$" dia. for the 2", $1\frac{1}{2}$", $1\frac{1}{4}$", and 1" and slightly smaller than the nominal sizes of the $3/4$" and $5/8$" ends. The two smaller ends are fitted with 2BA grub-screws and the larger ones with $1/4$" BSF grub-screws. The outer flange faces are relieved, leaving only a ridge around the edge about $1/8$" wide. It is most important that this face be finished finally by removing the chuck from the lathe and supporting each disc only by means of the stop bar in

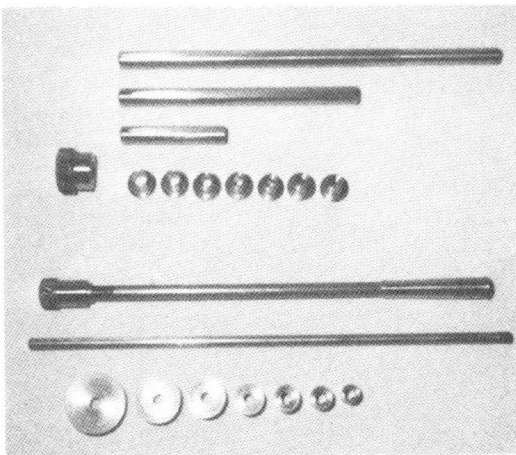

Three stop bars for length stop with supporting holder on left. Seven bushes to support long material through mandrel. Disc facing stop bar support with stop bar and six stop discs.

the taper sleeve and driven only from the grub-screw in both the disc and the knurled nut, having the boss of each disc just slightly clear of the taper sleeve. With a sharp facing tool and very light cut, a true face can be made on each disc, which, when the work is positioned against it in the chuck, will run true for facing the other side. The two grub-screws which bear on the stop bars should be ground flat on the points so as not to raise a burr on the flat of the stop bars.

CHAPTER 15

Automatic Facing and Boring Head

I have recently designed and made an Automatic Facing and Boring head which will bore and face to a maximum diameter of 6½" with an automatic facing feed of 0.005" per turn, which may be of interest to readers. In addition to these orthodox functions, the head may be used for flycutting, setting the diameter swept by the cutter to the most suitable size and traversing the job by means of the cross-slide. If suitable split bushes are made, the crank and cam journals on crankshafts can be machined. the throw being accurately set by means of the index plate on the feed screw of the head.

The usual orthodox facing feed is outwards from the centre, but a third hole is provided so that facing can be done towards the centre as for instance when it is necessary to face up to a shoulder. With the cutter bar in this position, turning a spigot on a part that is too big to swing in the lathe can be done, the job being held on the cross-slide or vertical slide.

I used the ³/₈" BSF thread provided in the Myford head for the collet attachment, for supporting a hardened steel roller against which the ten point star wheel impinges, this moves a ³/₈" BSF feed screw (20 t.p.i.) so giving the 0.005" per turn of the head. For boring, the roller is positioned clear of the star wheel and the slide is locked by means of the locking screw provided. I have made

The facing and boring head in position on the lathe.

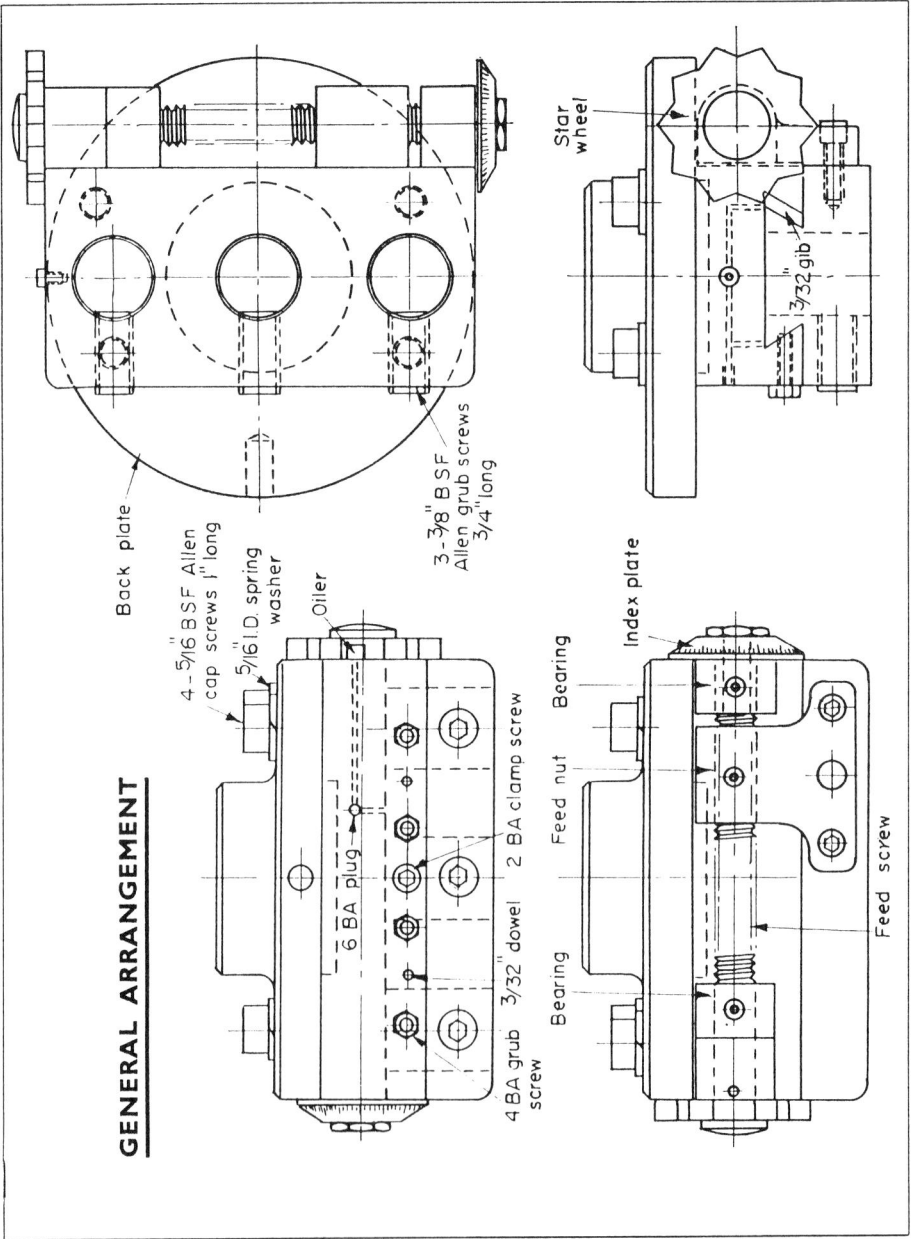

GENERAL ARRANGEMENT

Back plate

4 - 5/16" B S F Allen cap screws 1" long

5/16" I.D. spring washer

Oiler

6 BA plug

4 BA grub screw

3/32" dowel

2 BA clamp screw

3 - 3/8" B S F Allen grub screws 3/4" long

Star wheel

3/32" gib

Index plate

Bearing

Feed nut

Bearing

Feed screw

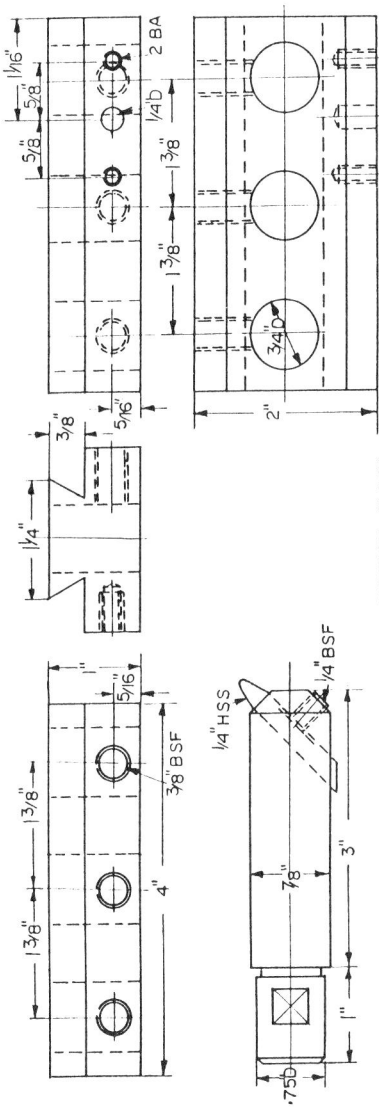

BOTTOM BLOCK
1 OFF   BMS

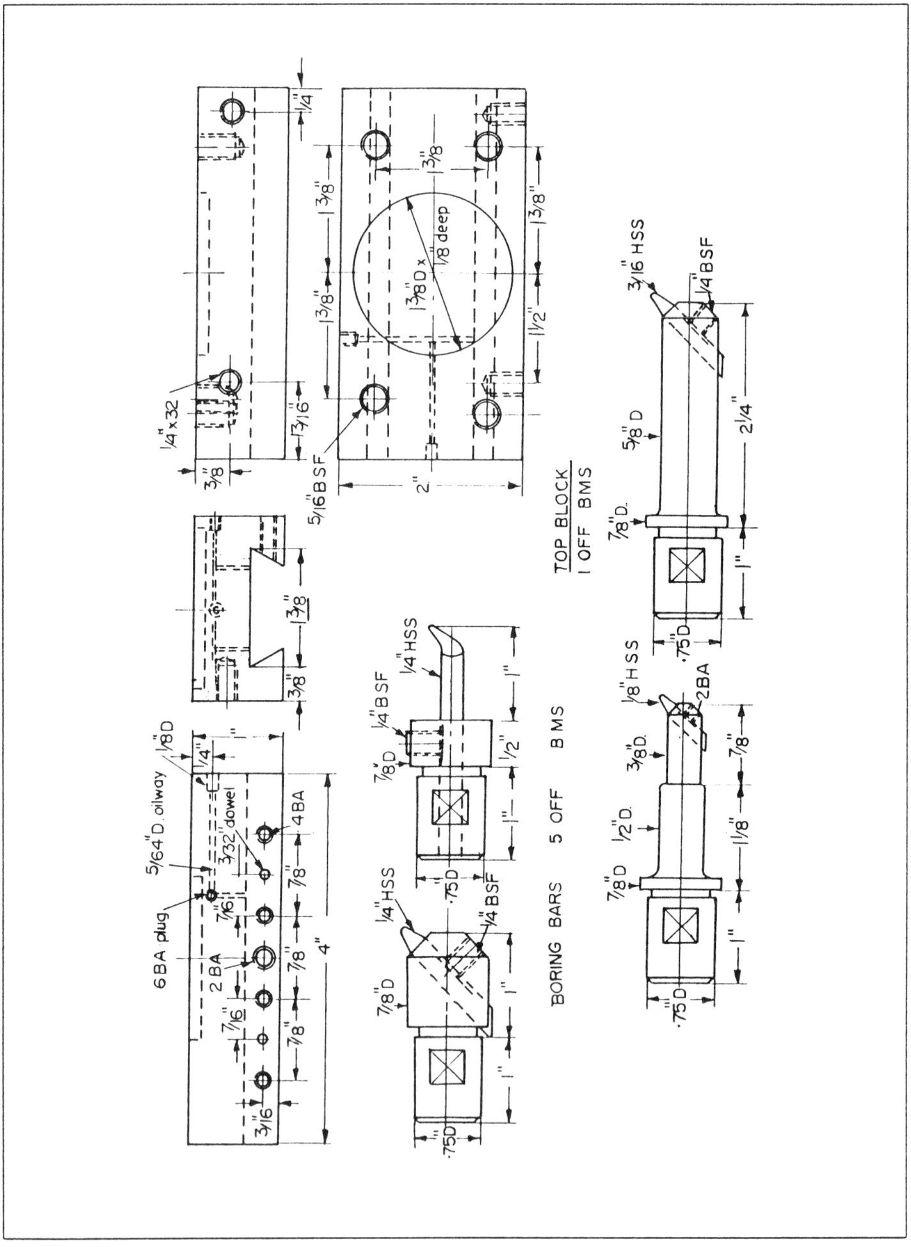
TOP BLOCK
1 OFF BMS
BORING BARS 5 OFF BMS
5/16" BSF
1/4" x 32
6 BA plug
2 BA
4 BA
3/32" dowel
5/64" D. oilway
1/8 D
1/4" HSS
1/4" BSF
3/16" HSS
1/8" HSS
2 BA
·75D

five boring and facing bars which take care of a wide range of work.

I had some 2" × 1" b.m.s. left over from the Quick Change Toolholders (discussed in Chapter 13) and I used this for slides, the backplate is a standard Myford 4" chuck backplate. At the maximum diameter, the head will clear the bed of the lathe, but in any case the total width is less than the width of the gap, so a swarf guard fitted to the saddle can be used, which is an important point.

Two pieces of the 2" × 1" steel were cut off 4" long full and faced off to 4". I found that these pieces were true on the faces to the surface plate and truly parallel, so decided not to face the pieces at all. The first piece was set up true to the dial gauge in the four-jaw chuck and truly central and bored 1³/₄" dia. to a depth of 0.125", undercutting slightly in the corner for the spigot to be turned on the backplate. The backplate was drilled ¹/₄" into the edge for a tommy bar, allowing for the turning off for the spigot, and turned on the face leaving the spigot at 0.100" depth a very close fit in the steel piece and also very slightly undercut. While working on the backplate it may as well be finished off; first get the outside diameter to 4", as you will find it slightly large and then mount the steel piece on the spigot at right-angles to the tommy bar hole and scribe two lines alongside the steel. Remove the steel piece and mark out the four holes for the ⁵/₁₆" Allen cap screws either Whit. or BSF. Note that one of the holes on the opposite side from the tommy bar hole is spread out ¹/₈" to clear one of the bearing piece spigot screws.

The steel piece can now be spotted through and drilled and tapping size, but do not under any circumstances go deeper than ⁹/₁₆" to the point of the drill. Make a flat bottom to the holes and tap with a plug tap to finish. Countersink all holes on both sides of the plate slightly and on the steel piece before tapping. Tap through the plate for truth. The steel piece is milled along the centre, first with an end mill and then with a 60 deg. dovetailed cutter. A ¹/₃₂" flat edge is left on the slides and on the tommy bar hole

side, this edge is ⁹/₁₆" from the centre and on the other side it is ⁷/₁₆" from the centre. This is to allow for the gib piece. This slideway must be truly parallel to the sides and exactly ³/₈" deep, and true to the face of the piece. Make sure of this. The other piece of 2" × 1" steel is also milled with an end mill and the same dovetailed cutter again, leaving the edges ¹/₃₂" wide and bringing the edges of the two pieces of steel flush. The gib piece is made of ³/₃₂" thick gauge steel which at the angle is about ¹/₈" wide, so a little latitude is allowed to get the sides flush while milling. Have a piece of gauge plate handy while milling and make it to a close push fit.

May I be allowed a little New Zealand 'skite' here? I did just this, and it is almost impossible to feel the edges at any position of the slides and the gib piece is quite a tight push fit. There are many ways of milling these slides, I have a 2" planer that I have converted to a plano-mill with No. 30 taper nose and double Timken bearings at the nose, ten speed back gear, four all-geared feeds and a lot of useful attachments, so the milling is now quick and no trouble. If I did not have this, I would probably have milled the slides in place on the nose of the lathe using my milling attachment, or it could be done on the vertical slide with the cutters in the chuck.

This second piece of steel is milled to 0.010" shallower than the other piece (0.365"), so that the bearing surface is on the outer two faces.

Some time ago I picked up from the foundry a 10" disc of cast-iron about ³/₄" thick. I faced this both sides of the lathe and used it as a lapping plate. I lapped both sides of the first slide and, with the gib piece removed, lapped the other slide into the first using fine paste. Next I clamped the gib piece in place with two clamps and a small piece of round steel and drilled the two ³/₃₂" dowel holes right through the gib at 1³/₄" centres and in the centre No. 24 for the 2BA Allen cap screw which is the clamping screw. The four gib screws are 4BA at ⁷/₈" centres and the four gib screw nuts are made from ¹/₄"

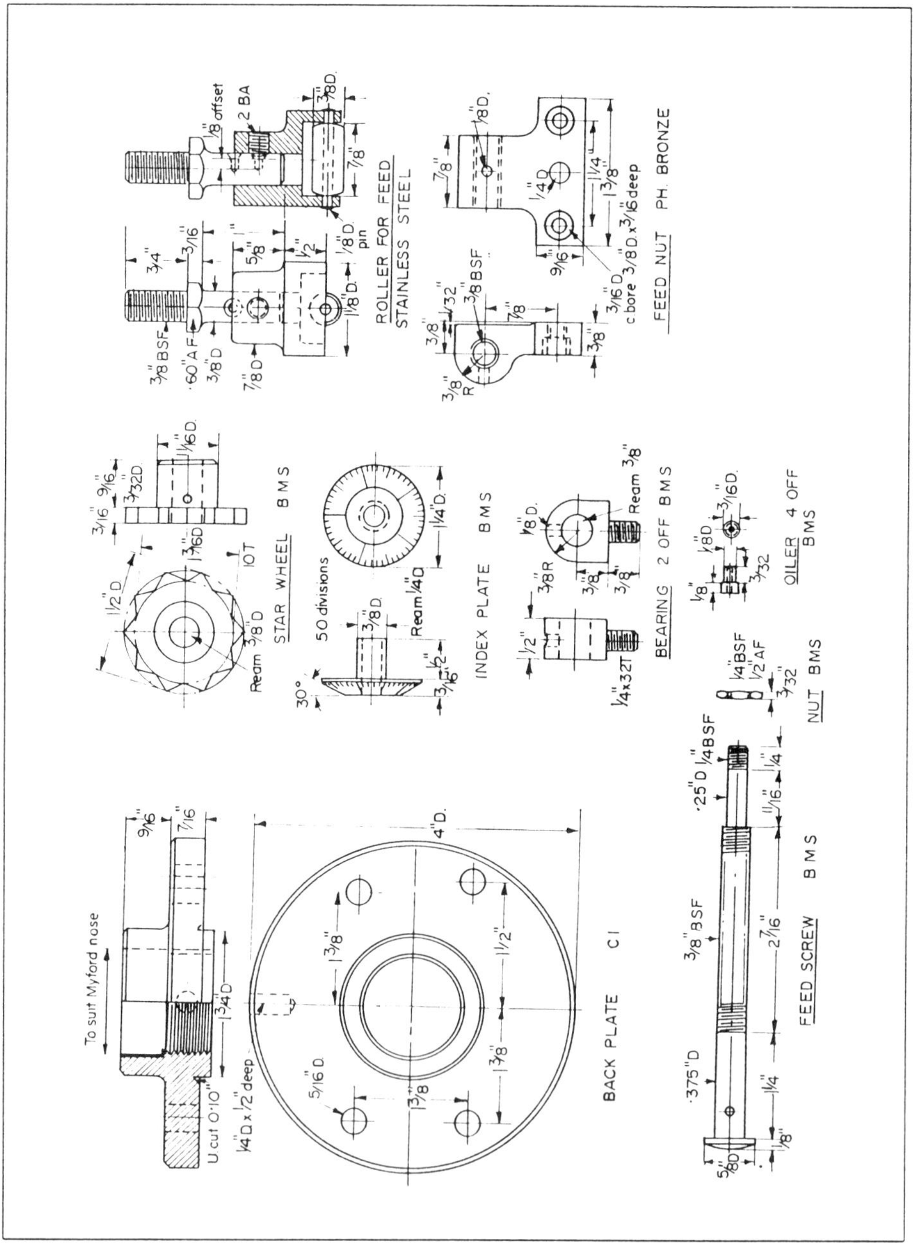
ROLLER FOR FEED
STAINLESS STEEL
FEED NUT PH. BRONZE
STAR WHEEL B M S
50 divisions
INDEX PLATE B M S
BEARING 2 OFF B M S
OILER 4 OFF BMS
NUT BMS
FEED SCREW B M S
To suit Myford nose
BACK PLATE C I

a.f. hexagon steel. These screws were all drilled until the point of the drill just indented slightly into the gib. The parts were all now cleaned up, the gib and screws all fitted and the backplate mounted using the cap screws cut to maximum length with $^5/_{16}$" spring washers under the heads, and the whole set up in the lathe with the slides flush on the ends and locking screw tightened up ready for boring and reaming for the bars.

The centre hole was centre drilled, drilled, bored and reamed $^3/_4$", allowing the boring tool to go below the surface of the first slide about $^1/_{32}$" to allow for reaming with the machine reamer; a hand reamer is useless for this job. Then setting the slide $1^3/_8$" off centre, using the vernier gauge, the second hole was bored and reamed and setting the opposite way the third hole was done the same way. The holes were very slightly chamfered on the outer edge. You will notice that the gib screws are on the same side as the tommy bar hole.

The three grub-screws for the boring bars are $^3/_8$" BSF, $^3/_4$" long and the points flattened off a little. These are drilled $^5/_{16}$" centres from the bottom edge and tapped.

As these slides are 'steel on steel', I provided for good lubrication by drilling a $^5/_{64}$" oil hole $1^1/_2$" deep and $^5/_{16}$" from the backplate in the centre of the end, the same end as the star wheel is fitted, and a cross hole to go through this drilled from the gib screw side and about $1^3/_4$" deep. Two vertical holes the same size were drilled from inside the backplate recess to just strike the corners of the slideways as shown on the drawing. The ends of these holes were plugged with a 6BA grub-screw except for the end one and four oiler nipples were made from $^3/_{16}$" b.m.s. turned a full $^1/_8$" for $^3/_{16}$" long, drilled through $^1/_{16}$" and parted off leaving a $^1/_8$" high head for the oil gun. One of these nipples was pressed into the end hole after drilling $^1/_8$".

Make a little pattern for the feed nut and get it away to the brass foundry for one off in phosphor-bronze, so that it will be ready when you need it. Now drill and tap the two $^1/_4$" × 32 t.p.i. holes no deeper than $^1/_2$" for the two bearing blocks. The one next to the star wheel is $^{13}/_{16}$" from the end and clears the offset hole drilled from the backplate, the other end is $^1/_4$" from the end. They are both a full $^3/_8$" from the backplate, and must be truly in line.

The two bearing blocks are made from $^3/_4$" key steel and this is set to run true in the four-jaw chuck and turned to $^1/_4$" for $^3/_8$" long and screwed with dies in the lathe $^1/_4$" × 32 t.p.i. The fine thread is necessary so that the blocks can more easily be set square. Part off the pieces square across at $^3/_4$" from the base and screw them into place, marking the outer face of each block. Setting this outer face $^1/_8$" out from the jaws of the chuck and with the centre $^3/_8$" from the base and truly central, $^1/_8$" is turned off, centred, drilled and reamed $^3/_8$". The other bearing block is machined exactly the same and then they are pushed over a stub mandrel and the opposite side faced off $^1/_8$" leaving the blocks $^1/_2$" thick × $^3/_4$" wide and $^3/_4$" high. The outer ends can now be rounded off to $^3/_8$" radius. To secure these bearing blocks permanently in place, I did not want to braze them with the possibility of distortion, so I spread a little Araldite over the threads and the bases and screwed them firmly into place getting the two bores truly in line with a piece of $^3/_8$" rod. This was left overnight until the star wheel and index plate were made.

A piece of $1^1/_2$" b.m.s. about 4" long was held in the chuck and the end faced and centred. A parting tool and left-hand turning tool was used to turn roughly down to $^3/_4$" for about $^3/_4$" long, leaving a $^3/_{16}$" wide disc on the end of the piece at the full size. A narrow side and face cutter was used and the dividing plate was set for ten divisions. The cut was taken to a depth of $^1/_4$" with the cutter set off centre $^{19}/_{32}$" to its inner face, so that when it was run around a second time with the cutter set $^{19}/_{32}$" from the other side of the centre, the distance between the two milled faces was $1^3/_{16}$". After milling and the points of the star very slightly rounded, the job was drilled, bored and reamed $^3/_8$" and parted off a full $^9/_{16}$" from the star.

The boring head in position on the lathe mandrel.

A stub mandrel was then turned and the journal turned to $^{11}/_{16}$" dia. and of a length that would allow for a $^1/_{64}$" fibre thrust washer between the bearing block, so that the star wheel just cleared the end of the slide; this should be about $^9/_{16}$" long.

The index plate was made from $1^1/_4$" b.m.s. and is turned down to $^3/_8$" a nice free fit in the bearing block for $^1/_2$" long, centred, drilled letter D and reamed $^1/_4$". It was parted off a full $^3/_{16}$" at the full dia., held in the $^3/_8$" collet and faced to $^3/_{16}$" and with the topslide set to 60 deg. it was turned until the edge was about $^1/_{32}$" wide. The graduating attachment was then set up at this angle and 50 divisions were indexed; the single divisions were made $^1/_{16}$" long, the fifth was $^1/_8$" long and every tenth was $^1/_4$" long. They were numbered with $^3/_{32}$" numbers in a clockwise direction. The nut which holds this index plate was now made from $^1/_2$" A/F hexagon steel and tapped $^1/_4$" BSF. The nut is $^3/_{32}$" thick.

We can now make the feedscrew to fit all these reamed holes and checking the lengths exactly from the job. The screw is made from $^5/_8$" b.m.s. and is made a tap-in fit in the star wheel, a free fit in the two bearing blocks and a tight push fit in the index plate. Another $^1/_{64}$" thick fibre thrust washer is fitted over the index plate and the shoulder on the feedscrew against which the index plate presses should be of such a length that the feedscrew will be free but without end play. The index plate is fitted this way so that it can be set to coincide with the zero mark on the fixed slide on the tenth and fifth divisions each turn of the mandrel. The thread at $2^7/_{16}$" long is $^3/_8$" BSF and must be screw-cut to fit an item which is of phosphor-bronze.

The phosphor-bronze nut should now be machined on the back face, noting that it is relieved to clear the fixed slide by $^1/_{32}$" and to clear the backplate. The base should well clear the outer face of the moving slide and clear the fixed slide. It is drilled No. 24 at $1^1/_4$" centres and centred in the centre only, for the dowel pin. It is now clamped to the moving slide with the centre at $1^1/_{16}$" from the same end as the index plate. This will leave $^1/_8$" of further offset movement to clear a boring tool more easily.

The moving slide is now drilled No. 24 through from the nut to $1/2$" depth to the point of the drill, and drilled letter D for the dowel pin, which is reamed $1/4$" and a silver steel dowel fitted. The nut is counterbored $3/16$" for the two $1/2$" × 2BA cap screws and the heads are set in so that they are just slightly above flush.

The two slides together are next set up on the cross-slide of the lathe with a true running piece of $3/8$" rod in the chuck and the back centre which is passed through the bearings. When clamped to the cross-slide, this rod should still be free to turn and the cross-slide is locked. This is to bring the two bearing blocks in line with the centres of the lathe. The feed nut is mounted complete with screws and dowel pin and a sharp $3/8$" drill is put in the three-jaw chuck and the nut is drilled until the lip of the drill just enters. The drill is now changed for a $21/64$" drill and the nut drilled through. The nut can be tapped through the bearing, bringing the plain shank of the tap to line up in the bearing. This will make a truly in-line thread which will run very freely.

Star wheel

The star wheel is pinned with a $3/32$" silver steel pin, but after drilling and before the pin is fitted, the star wheel should be thoroughly case hardened. The spigot for the roller holder is made from a piece of 0.600" hexagon steel (Whit. size) and is first turned to $3/8$" for 1" long, leaving a nice radius and then reversed in the chuck and turned $3/8$" leaving a full $3/16$" of hexagon. This is now screw-cut and threaded to fit the Myford $3/8$" BSF thread in the head of the lathe. Check the position of this hole by fitting the spigot and the facing head. If it is the same as mine, there will be a full $1/8$" between the edge of the backplate and the $3/8$" shank of the spigot, and this will allow the roller holder to be removed at any time without removing the facing head.

There is no reason why this eccentric roller holder should not be turned in the facing head, using the graduations to ensure the correct offset. So cut off a piece of $11/4$" b.m.s. $21/2$" long and turn one end a close fit in the $3/4$" reamed holes of the head to 1" long. File a flat for the grubscrew (these flats for the boring bars, by the way, are filed slightly deeper on the end nearest the shoulder, which prevents them working out). This is now mounted in the central hole and set to run true to the gauge and the dial is set to the zero mark. Set the slide $1/8$" eccentric and lock it and turn to $7/8$" dia. for a length of $5/8$". Now set out a further amount to $1/4$" eccentric and turn to $11/8$" dia. to close to the slide and centre the end, drill, bore to clean up and ream to $3/8$" to fit the spigot. This should leave the thin side approximately $1/8$" thick. The piece should now be parted off at $11/4$" long.

Gripping the $11/8$" dia. part in the three-jaw chuck with parted off face outward, the end is now faced and recessed to $7/8$" dia. to a depth that will leave $1/8$" of metal at the bottom of the recess. The edges of the recessed part are cut away, leaving the two lugs for the roller pin $3/16$" high and in line with the offset. The roller is turned to $3/8$" dia. and the centre hole is reamed $1/8$", the ends of the roller being radiused to fit in the $7/8$" dia. recess. The roller and pin are made from silver steel and hardened. The $1/4$" BSF grub-screw is fitted in the centre of the $7/8$" dia. part right on the high side. To get the position of the two holes centre drilled in the spigot, one for engagement of the roller and the other for clearance, the spigot is tightened up in place and the roller is put in place, the facing head fitted and the roller set so that it will just clear the index plate, the angle of the roller is then set so that the star wheel strikes it squarely. The grub-screw is tightened in this position and tightened again in a position $1/4$" further in to clear the star wheel. The indentations made by the grub-screw can now be drilled with a centre drill to about $1/4$" wide in these two positions.

Apart from the fitting of the other three oiler nipples, one in each of the bearing blocks and the other in the centre of the feed nut, all that remains now is to make the five

Some of the components of the boring head showing the tools used.

boring and facing bars. They are all made from $^7/_8$" b.m.s. Cut off one piece $4^1/_8$" long, one piece $3^5/_{16}$", another $3^1/_{16}$" and one piece $3^3/_4$" long, which will make the two shorter bars. I set up each piece in turn in the three-jaw chuck and faced and centred the end. With a parting tool $^1/_{16}$" thick, I cut into the steel 0.070" at 1" long to the inner face of the tool and turned to a very close fit in the bores of the facing head, bevelling the outer corner slightly. This was made a good twist-in fit in each case, not forgetting to do both ends of the $3^3/_4$" long piece. The flats were all filed now, using a good square file and making them $^3/_8$" square, but as stated, slightly deeper on the shoulder end to give a slope to the bottom face. Getting the centre hole in the facing head true to the dial, the slide was locked and each tool was faced

on the end, bevelled and turned to the drawing. The bevelled ends are at 45 deg. and are made the same width as the tools as shown. The tools are round tool bits in h.s.s. with a flat ground on the side for the Allen grub-screws. The smallest tool is a piece of $^1/_4$" round which had to be forged over, ground to shape and then re-hardened. The grub-screw in this case is at right-angles.

The tools were all ground with slight top rake (positive) and also a slight positive side rake and with ample clearances on the end and the diameters. They bore and face to a beautiful finish and accurate diameters. This facing and boring head can be made in a few hours and amply returns the time to make it over makeshift methods which are seldom satisfactory.

CHAPTER 16

A Worm-Wheel Hobbing Attachment

Worm drives are very useful mechanisms which the home workshop enthusiast is often called upon to make, whether as a speed reduction drive for various purposes or as a ready means of indexing. However, it is impossible to produce an accurate worm-wheel for such things as indexing by hobbing in a free state, even after preliminary gashing, as the pitch circle diameter is being reduced as the hob is fed into the wheel resulting in teeth of varying pitch and thickness. Having been troubled by this problem for many years and finding it impossible to rig up any sort of makeshift drive to the worm-wheel blank, I decided to design a proper fixture that would take care of a wide range of worm drives and produce a truly accurate worm-wheel for any indexing need without the tedious need to gash first, or if necessary, to produce a worm-wheel without the need to make a hob.

This I have done and I hope it will interest a great number of readers. One photograph shows a 30 T., 15 D.P. worm in bronze that was hobbed directly into the blank without preliminary gashing in less than 20 seconds. It proved extremely rigid in operation, without any sign of vibration or snatching and I am confident that it is extremely accurate.

Another photograph is a cast iron worm-wheel of 60 T. but it is a 5 start giving 1:12 reduction, and was cut entirely without a hob using an ordinary boring bar and a single cutter of 29 deg. included angle, set to the helix angle and the diameter of the worm. The 5 starts are obtained using my 12 and 10 ring indexing catchplate used mainly for multiple start threads. It is important when making a worm-wheel without a hob that the diameter set by the cutter point is correct, as this determines not only the correct radius of the teeth, but also

The complete hobbing fixture using a single point tool.

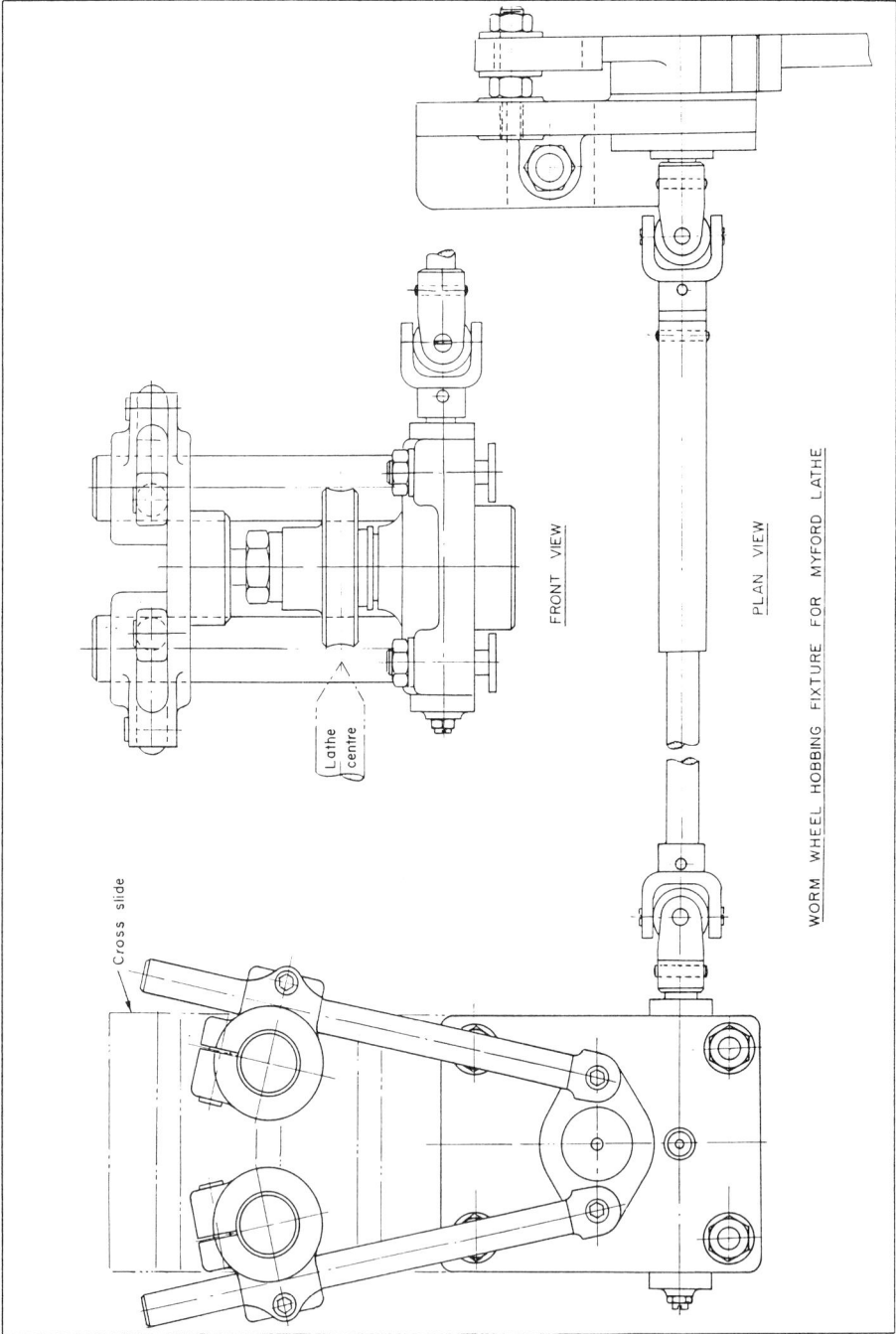

FRONT VIEW

PLAN VIEW

Lathe centre

Cross slide

WORM WHEEL HOBBING FIXTURE FOR MYFORD LATHE

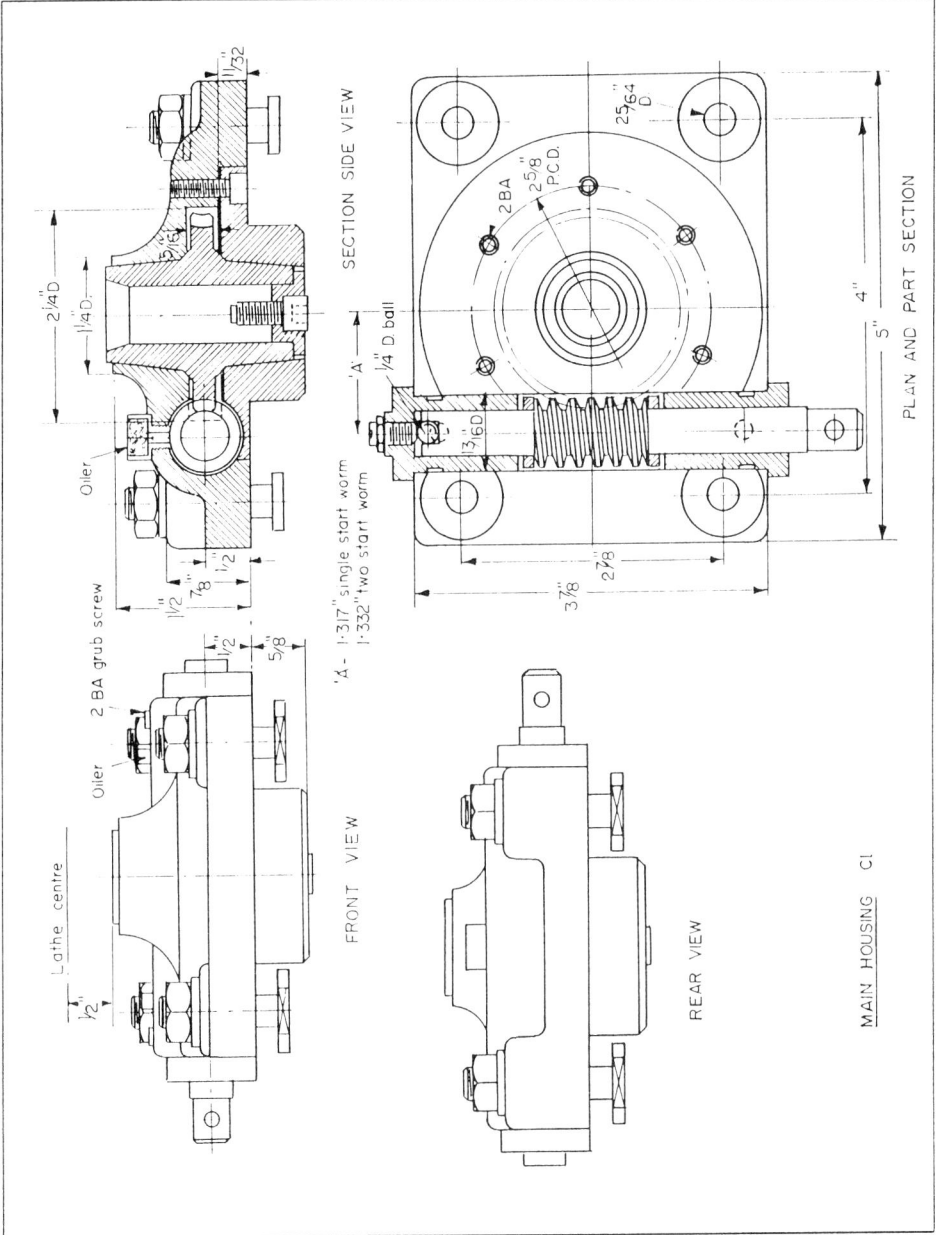

SECTION SIDE VIEW

2 5/64" D

2 BA

2 5/8" PCD.

1/4" D ball

'A'

'A' - 1·317" single start worm
1·332" two start worm

3/16 D

2 7/8"

3 7/8"

4"

5"

PLAN AND PART SECTION

2 1/4 D

1 1/4 D

Oiler

1 1/32

7/8"

1/2"

Oiler

2 BA grub screw

1/2"

5/8"

Lathe centre

1/2"

FRONT VIEW

REAR VIEW

MAIN HOUSING   CI

Hobbing a 60 tooth, 5 start worm wheel using a single point tool. Note the indexing catchplate.

the correct helix angle. The straight sided cutter also produces the correct involute shape of the teeth.

In designing this fixture, I decided to use the screw-cutting change wheels to determine the number of teeth and the tumbler gears to determine whether right or left handed, and I decided that the number of teeth would be twice that as set for screw-cutting. Any gaps in the range could be corrected by using a second set of change wheels driven by the tailstock end of the leadscrew and driving a splined and universal jointed shaft back to the worm drive in the fixture. This is shown in another photograph. The greatest diameter of worm-wheel that can be hobbed is more than 12" and the smallest is down to nil; the number of teeth can be anything at all, so the fixture could be said to be truly universal.

The second set of change wheels is used in the following manner. Say 30 teeth are required. In the Myford gearbox there is no position for 15, 30 or 60 teeth, so we set for 20 teeth in the gearbox which with one-to-one gearing would give us 40 teeth on the worm gear blank. We have to speed up the blank to cut 30 teeth in the ratio of 40-30,

Hobbing a 30 tooth worm wheel direct from the hob, time 20 seconds.

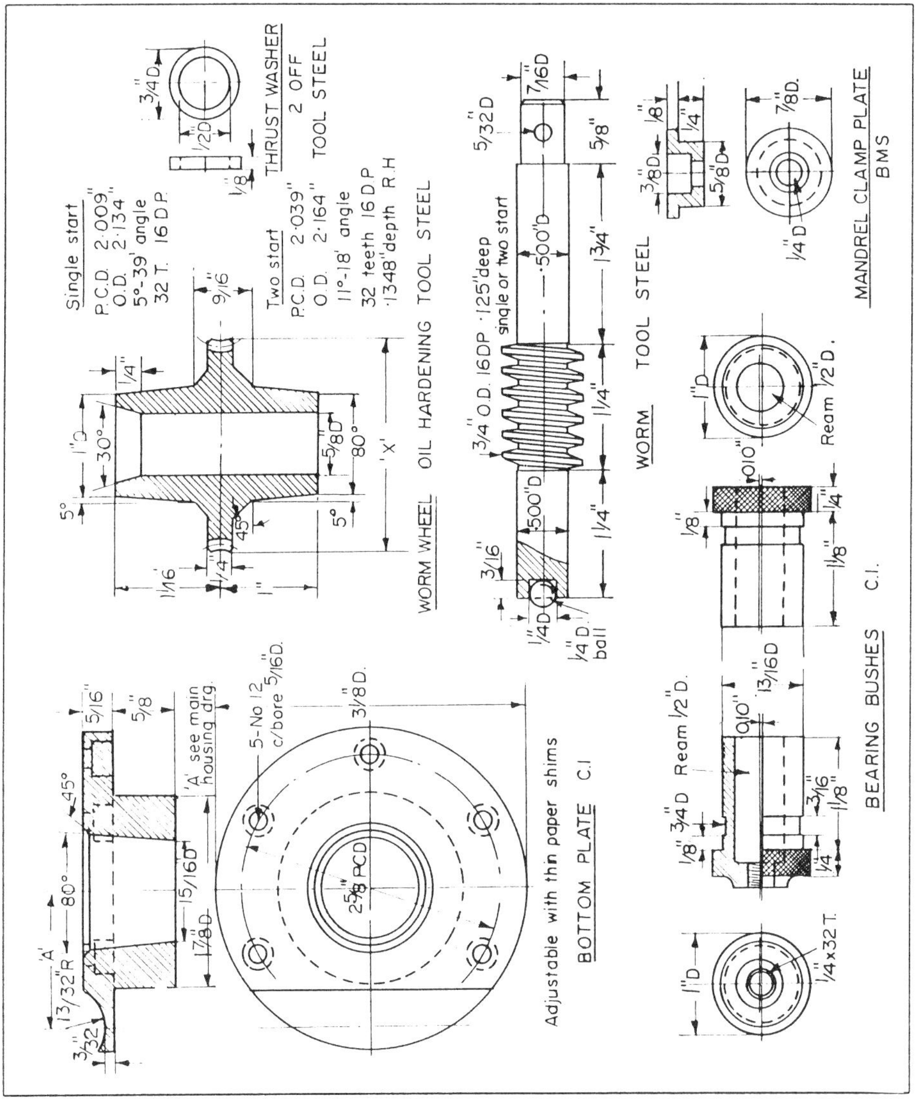

Single start
P.C.D. 2·009"
O.D. 2·134"
5°-39' angle
32 T. 16 D.P.

Two start
P.C.D. 2·039"
O.D. 2·164"
11°-18' angle
32 teeth 16 D.P.
·1348" depth R.H

WORM WHEEL OIL HARDENING TOOL STEEL

THRUST WASHER
2 OFF TOOL STEEL

WORM TOOL STEEL

3/4 O.D. 16 DP ·125" deep
single or two start

MANDREL CLAMP PLATE
BMS

Ream 1/2 D.

BEARING BUSHES C.I.

BOTTOM PLATE C.I

Adjustable with thin paper shims

5-No 12
c/bore 5/16 D.

2⅝" PCD

'A' see main
housing drg.

1/4 × 32 T.

so we can fit a 40 T wheel on the end of the leadscrew and a 30 T wheel on the spline shaft drive (or 60-45) so that a gear wheel of the same number of teeth as the change gears are set for (×2) on the leadscrew end,

and a change gear of the required number of teeth on the blank, will give us the correct gear ratio. Failing a gear wheel of the right number of teeth, this can be obtained by compound gearing at the tailstock gear train

153

The quadrant at the tailstock end of the lathe.

or in conjunction with a compound train at the screwcutting train of gears.

The fixture is bolted to the cross-slide and no provision is made for height adjustment, but worm-wheels ³/₄" wide can be hobbed against the shoulder of the mandrel; thinner wheels are provided with shoulders or spacing washers. The top of the mandrel is well supported by a ball race in a housing, which is supported by two 1" dia. vertical columns rigidly bolted to the cross-slide, which connect to the ball race housing by two ¹/₂" adjustable bars. The driving worm-wheel built into the fixture is integral with its double tapered shafts and is hollow to support the mandrel or collets. It is made of oil-hardening tool steel and is adjustable in its bearings for wear. The worm is adjustable in eccentric phosphor-bronze bearings and is also of oil-hardening tool steel; the teeth are 16 D.P. 2 start and the worm-wheel of 32 teeth, which is narrow, is milled at the helix angle of the worm, but is not hobbed as this would destroy its accuracy.

Now the first things to make are the six simple patterns. The main housing consists of two patterns both with ³/₄" core prints

each side, the mandrel support bearing flange also has a ³/₄" core print each side and the pattern for the two bar clamp castings also has the same size core print each side. The quadrant casting pattern has a 1¹/₄" print each side and the bracket for this quadrant has a ⁵/₈" core print each side for the drive spindle bushes. I had intended to make my quadrant with core prints for the ³/₈" wide slots and make three core boxes for these slots, but the temperature on the day I made this pattern was in the mid-nineties and I could not rake up sufficient energy to make three core boxes, so left the pattern flat each side and milled out the slots when the castings were machined; there are no other core boxes to make. All the core prints are ⁵/₈" high.

A piece of 2¹/₄" dia. oil-hardening tool steel for the worm-wheel will be required 2¹/₄" long and a piece of the same steel ⁷/₈" or 1" dia. and 5" long will be required for the worm. I highly recommend that the worm be made in the two start version if at all possible or else for many worm-wheels the spline shaft will have to be speeded up 4-1 instead of 2-1 at the most, and a greater

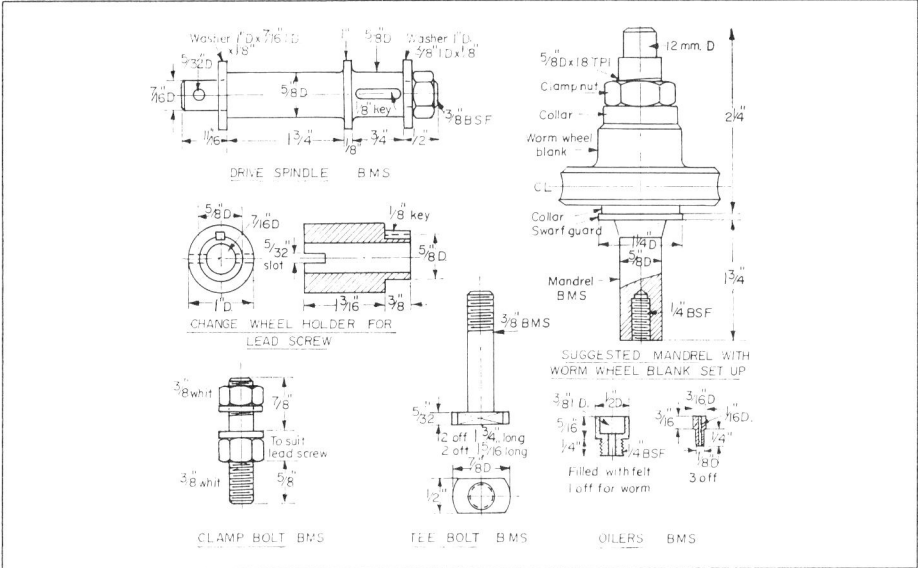

Washer 1"D x 7/16"D x 1/8"   5/8"D   Washer 1"D, 7/8"ID x 1/8"

5/32"D
7/16"D   5/8"D   1/8" key   3/8"BSF
1/16"   1 3/4"   3/4"   1/2"

DRIVE SPINDLE   B.M.S.

5/8"D x 18 TPI   12 mm. D
Clamp nut
Collar
Worm wheel blank
C.L.
2 1/4"

5/8"D   7/16"D   1/8" key   Collar
5/32" slot   5/8"D   Swarf guard   1 1/4"D   5/8"D   1 3/4"
1"D   1 7/16"   3/8"   Mandrel B.M.S.   1/4" BSF

CHANGE WHEEL HOLDER FOR LEAD SCREW

3/8" BMS

SUGGESTED MANDREL WITH WORM WHEEL BLANK SET UP

3/8" whit   7/8"
To suit lead screw
3/8" whit   5/8"

CLAMP BOLT   BMS

5/32"
2 off 1 3/4" long
2 off 1 5/16" long
5/8"D
1/2"

TEE BOLT   B.M.S.

3/8"I.D.   1/2"D   3/16"D
5/16"   3/16"   1/16"D
1/4"   1/4"BSF   1/4"
Filled with felt   1/8"D
1 off for worm   3 off

OILERS   B.M.S.

selection of gear teeth can be cut. For those who have made or are making my speed reduction head, set up for 48 D.P. for cutting the worm, which is quite a fine gear train, without this head it will be necessary to set up for 8 D.P. which is very coarse. (The single start version is 16 D.P.) The gearing for 8 D.P. is 55 T on the spigot and 35 T on the leadscrew with a step up of 2-1 between. It may be possible to drive the lathe by hand using the leadscrew hand-wheel, or take it very carefully in slowest back gear.

The worm-wheel is bored and reamed 5/8"

End milling the 3/8" curved slot.

155

after facing in the 4-jaw chuck and the end is tapered 30 deg. included, a $1/4$" wide taper being required to centralise the hobbing mandrel or for the possible use of collets. The design of the bearing portions of this worm-wheel is the same as was once used in precision lathe spindles and can be adjusted for wear by taking out shims from the bottom flange of the main casting.

The lower part of the worm-wheel has its 45 deg. part $3/16$" wide as this takes most of the thrust, while the upper part has its 45 deg. part only $1/8$" wide. The main journal parts are 5 deg. a side or 10 deg. included, the smallest diameter at the ends being 1". This design is made for the Super 7 and perhaps it may be necessary to reduce the bottom diameter for the ML7. The 32 teeth 16 D.P. were milled at 11 deg. 18 min. using the taper turning attachment and my milling attachment (Chapter 1), the O.D. being 2.039", depth 0.1348", and the teeth cut right hand. Do not harden the worm-wheel or worm until the housing is machined, as it will be necessary to trim off each end of the journals, the top to stand just proud of the machined casting and the bottom to be $1/8$" short to allow for the mandrel clamp.

The housing is machined for the worm-wheel by holding in the 4-jaw chuck, after trueing up the top, to enter into the body of the chuck; grip on the edge of the round raised part. It is faced on the bottom, recessed $11/32$" deep to a diameter of $3^1/8$", a further recess is bored $5/16$" deep to a diameter of $2^1/4$" and is then taper bored to fit the worm-wheel so that the bottom edge of the $1/4$" wide teeth comes flush with the

bottom of the inner recess. The best way to get a good fit on both tapers is to set the taper turning attachment to 5 deg. and the topslide to 45 deg. and work them a bit at a time until marking with blue shows up on both tapers at the right depth.

The bottom flange is turned to fit the topslide hole in the cross-slide of the lathe and to fit the $3^1/8$" dia. recess in the housing casting. It is then set true to the dial gauge in the 4-jaw chuck for facing and boring. The

QUADRANT C.I.

BRACKET FOR QUADRANT C.I.

flange is faced to $5/16$" wide only to allow for shims or perhaps slightly less and is then bored to fit the worm-wheel in the same manner as the housing casting. When correctly bored, the bottom flange will be very slightly inside flush with the housing. The five 2BA screws are now spaced out for 6 positions at $1^5/16$" radius using the indexing attachment and drilling through a drill jig set up in the tool-holder. I turned a little off the underside of the five 2BA Allen screw heads for greater strength under the heads.

The shims are only thin paper and are quite satisfactory, the quantity is determined after hardening and running in the worm-wheel in its journals. The hardening of the worm-wheel is done by bringing to a bright red heat and plunging into oil; the tempering is done by bringing the teeth to a very dark straw colour slowly and so allowing a little colour to creep through the rest of the piece which is thus left almost dead hard. The worm is finished off for fit with the worm-wheel by hobbing the worm-wheel into the worm, setting up the worm-wheel with its

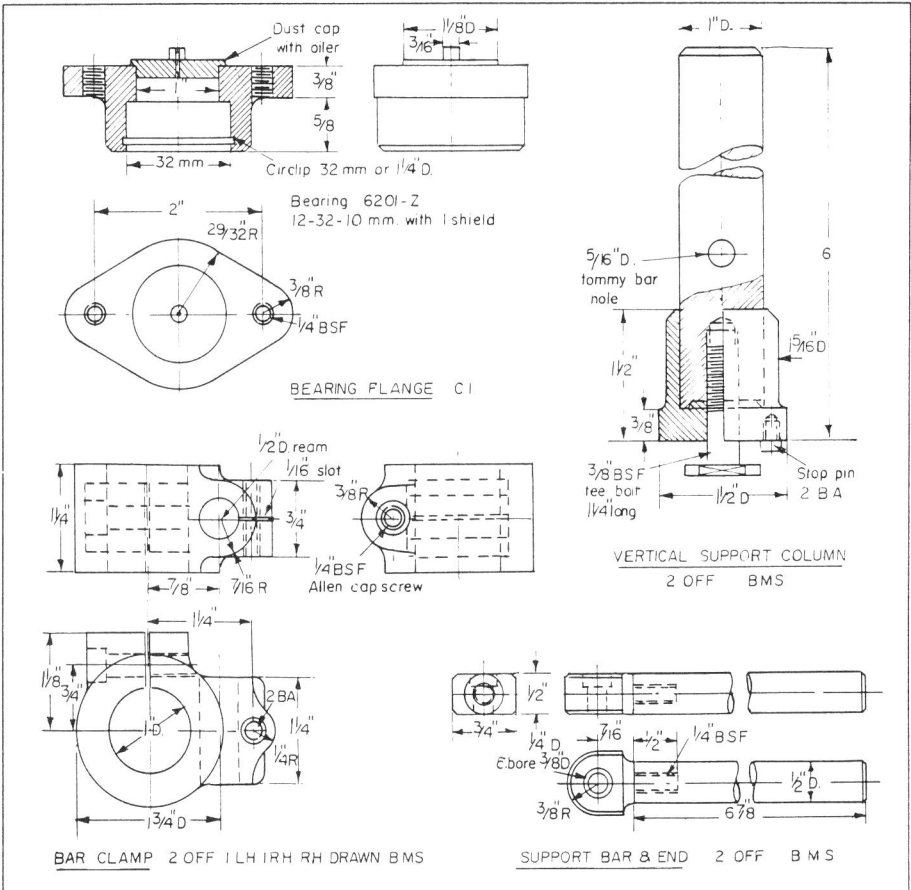

Dust cap with oiler

$1^1/8$ D

$3/16$

$3/8$"

$5/8$

32 mm

Circlip 32 mm or $1/4$ D.

Bearing 6201-Z 12-32-10 mm with 1 shield

2"

$29/32$ R

$3/8$ R

$1/4$ BSF

BEARING FLANGE C.I.

1"D.

$5/16$"D. tommy bar hole

6

$5/16$ D

$1^1/2$"

$3/8$

$3/8$ BSF tee bolt $1/4$ long

$1/2$ D

2 BA

Stop pin

VERTICAL SUPPORT COLUMN 2 OFF BMS

$1/2$ D. ream

$1/16$ slot

$3/8$ R

$3/4$

$1/4$ BSF Allen cap screw

$1^1/4$"

$7/8$

$7/16$ R

$1^1/4$

$1/8$"

$3/4$

1 D

2 BA

$1^1/4$

$1/4$ R

$1^3/4$ D

BAR CLAMP 2 OFF 1 LH 1 RH RH DRAWN BMS

$1/2$"

$3/4$

$1/4$ D. C.bore $3/8$ D

$3/8$ R

$7/16$

$1/2$

$1/4$ BSF

$1/2$ D

$6^7/8$

SUPPORT BAR & END 2 OFF BMS

A WORM-WHEEL HOBBING ATTACHMENT

Cutting the 2 start 16 D.P. worm using the speed reduction head.

edge to the centre line of the worm. This is done until a perfect fit is obtained and the centre distance is 1.332" by measuring over the outside of both worm and wheel which should measure 2.789". The worm can now be hardened, plunging vertically into oil for quenching, tempering to a medium straw colour.

Boring for the worm bushes is done by means of a drill and a boring bar while bolted to the cross-slide in the following way. Fit the worm-wheel into place in its housing with the bottom plate tightened up and fit into place in the cross-slide hole but in the reverse position, that is with the front edge to the rear. Set a stop block to touch the centre of the edge at the rear; I clamped a small square in the centre. Now fit a true mandrel in the worm-wheel and set the cross-slide to the centre of the mandrel using a wobbler; do not forget to allow for the thickness of the wobbler. Having found the centre of the mandrel, move the cross-slide the centre-to-centre distance of 1.332" plus 0.010" using the cross-slide gradua-tions. The extra 0.010" is to allow for half of the eccentricity of the worm bushes. Having now got the correct lateral position for boring, it is necessary to pack up to the correct height to bring the centre of the hole $1/2$" above the bottom face; this took $15/16$" of

packing in my case and the housing was brought once again in contact with the stop block or small square still clamped to the cross-slide in the centre and so this restored it to the correct centre distance for boring. Clamping securely in this position, the face was centred, a small drill was entered well and then drilled right through $23/32$", after removing the hardened worm-wheel of course and tightening the bottom flange close up against the main housing to give full clearance to the bottom flange. A boring bar was now used to bore to $13/16$" for the eccentric bushes.

The bushes of either cast iron or phosphor bronze are first turned on the outside using a little extra in each case to hold in the four-jaw chuck and are then set over to show a difference of 0.020" reading on the dial gauge and bored and reamed $1/2$" to fit the worm shaft. The left-hand one is a blind hole and drilled and tapped 1" × 32 tpi or 26 tpi for a thrust screw which is hardened and bears against a $1/4$" steel ball for thrust adjustment. The grooves turned on the outside of the bushes are for the grub screws which hold the eccentric adjustment, a $1/8$" tommy bar hole is drilled in the flange (not shown on the drawing) for adjusting the bushes.

Two collars or thrust washers, in

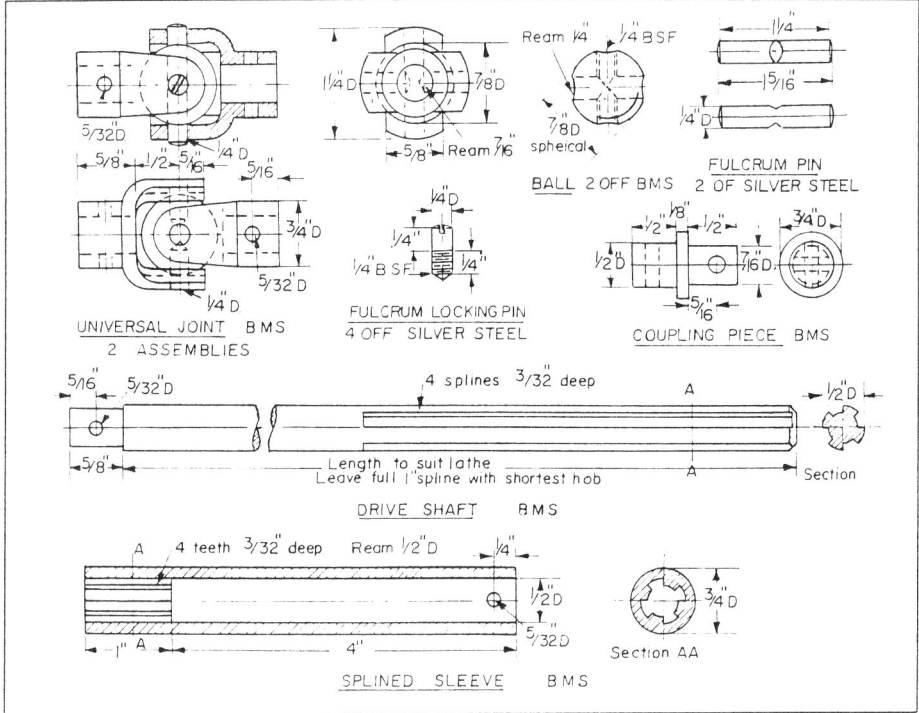

hardened tool steel $3/4" \times 1/2" \times 1/8"$ thick, fit each side of the worm to bear against the edges of the bushes; these are also hardened. These parts can now be assembled, not forgetting the two small oilers and the large one which is filled with wool, the wool passing through the hole and knotted each end of the hole; it can be run in using the lathe and adjusted to close but free running. The tee bolt holes are marked from underneath, drilled and spot faced so that the tops of the near two are slightly below the level of the flange of the mandrel when made, this to clear a large diameter worm-wheel that may be hobbed.

The quadrant bracket casting is first machined on the edge to give a true face for setting up for machining the bottom face, and should not be wider than $1 5/8"$ to clear the gap in the lathe for turning. I was able to machine the bottom leaving the $1/4"$ thick centre portion to fit the gap in the lathe bed, but if this face is machined on the cross-slide using a facing cutter, it will be necessary to machine the whole face off flat, afterwards fitting a piece of $1/4"$ thick steel plate $1 3/8"$ wide in position by dowelling and screwing. The clamp plate is $1 3/4"$ square by $3/8"$ thick and tapped for the clamping bolt which I made $7/16"$ B.S.F. The casting can now be set up on the faceplate for turning and boring and facing. I had to drill and tap a $3/8"$ hole in a strategic place to assist in holding the casting securely. The boss is turned to $1 3/8"$ dia., faced on the outer and inner faces and bored and reamed $7/8"$ for the drive spindle bushes and the face for the clamping bolt is faced at the same time.

The quadrant will just clear the gap and can be held to the faceplate by drilling and

Milling the worm teeth for the fixture using the taper turning attachment, etc.

tapping holes which will later be cut out when slotted. It is faced both sides to $^{15}/_{32}$" thick and the curved $^{3}/_{8}$" slot for the clamp bolt was end milled while set up on the faceplate, using my milling attachment, the drive being taken by the indexing attachment (Chapter 2). It is bored a working fit for the bracket and the position of the clamp bolt in the bracket can be marked off from the quadrant. It is fitted with a $^{1}/_{4}$" B.S.F. clamp screw and slit $^{1}/_{16}$" for clamping.

The clamp bolt is made from 0.600" hexagon steel screwed to fit the $^{3}/_{8}$" tapped hole in the bracket and is fitted with a washer to go each side of the quadrant and a $^{3}/_{8}$" Whit. nut. The hexagon part is made the proper length to suit.

The drive spindle is made from b.m.s. and will take two change wheels clamped against the integral shoulder by a washer and $^{3}/_{8}$" B.S.F. nut; a washer at the other end takes up end play when the universal joint is fitted.

The change wheel holder which fits on the end of the leadscrew of the lathe in place of the handwheel is made of such a length that the change wheels will just clear the end of the lathe bed. It duplicates the bore and slot of the hand-wheel exactly and is held in place by a thin washer and the hand-wheel nut. It is fitted with a $^{1}/_{8}$" key and is the width of one change wheel.

The two universal joints comprise two $^{7}/_{8}$" dia. spheres turned from $^{7}/_{8}$" rod using my spherical turning attachment (Chapter 5) and are bored and reamed through the centre after turning to $^{1}/_{4}$" for the $^{1}/_{4}$" silver steel pins, and are then drilled and tapped $^{1}/_{4}$" for the $^{1}/_{4}$" silver steel pins, and are finally drilled and tapped $^{1}/_{4}$" b.s.f. for the locking pins at right angles. The four swivelling parts of the universals are made of mild steel $1^{1}/_{4}$" dia. and are bored a close but free fit for the $^7/_8$" dia. spheres to the drawing and are reamed $^7/_{16}$" to fit the shafts and drilled $^5/_{32}$" for the pins. The holes for the fulcrum pins are reamed $^1/_4$" and must be truly central, by indexing.

After drilling etc., the unwanted portions were cut away by end milling and finished off by filing. The shaft splined sleeve is made from $^3/_4$" b.m.s. and is drilled and reamed $^1/_2$" for 4" long, the remaining 1" is drilled $^{13}/_{32}$" and is slotted with four slots $^1/_8$" wide to a full $^3/_{32}$" deep using my slotting attachment (Chapter 6) and the dividing attachment. It is finally fitted with a coupling piece shouldered in the middle to fit the splined sleeve and the universal joint and pinned to both with a $^5/_{32}$" pin in each. The splined shaft must be made so that there is a full 1" of spline in the splined sleeve when the main housing will just clear the chuck of the lathe in a position to suit the shortest hob that can be used; the splines on the shaft should be a full 5" long and will thus give ample movement to suit any hob.

The splines

The splines were cut using a $^3/_{32}$" wide cutter set to the edge at $^1/_{16}$" off centre and four cuts made, then the cutter was set $^1/_{16}$" to the other side of the centre line and cut through again. The depth for all the cuts was the same at a full $^3/_{32}$". The spline does not depend on the depth of the cut for true running but rides on the sides of the teeth. It should be made free sliding, but without slack. The shaft is made of $^1/_2$" b.m.s.

The mandrel support bearing flange casting in cast iron is bored a tap fit for the ball race which is a 6201Z and has one shield which goes to the bottom, it is held in place with a circlip of 32mm or 1$^1/_4$" of the inside fitting type. Above the bearing it is bored 1" and it is fitted with a plate to keep out dirt, with an integral oiler turned and drilled $^1/_{16}$". The top face is faced off true and drilled and tapped $^1/_4$" b.s.f. at 2" centres for the bar support screws. I made my bar support ends of $^3/_4$" round b.m.s., and spherically turned them and then milled

off just proud of the $^1/_2$" shafts before drilling; but they can just as easily be made of $^3/_4$" × $^1/_2$" b.m.s. if you have it. The bar clamps are both made from the one pattern, bored a slide fit on the 1" shafting, fitted with $^1/_4$" Allen clamp screws and slit for the 1" shafts and fitted with 2BA Allen clamp screws for the $^1/_2$" shafts and split here also. Take extra care to get the 1" and $^1/_2$" shafts at truly right angles. They should be quite free on both shafts when the clamp screws are loosened for easy setting. The 1" shafts, two off, are about 6" long and fit into b.m.s. bases made from 1$^1/_2$" shafting. They should be a close fit, but free to turn. The bases are fitted with a 2BA stop screw to prevent the bases from turning and marking the cross-slide when the bars are tightened up. Tightening up is done by using a $^5/_{16}$" tommy bar in a $^5/_{16}$" hole drilled in each bar. A quarter turn to slacken and the whole bar is slid out of the slot to remove. The bottoms of the 1" bars are made truly flat and undercut as shown, but the ends of the bases should be only very slightly undercut.

I always make my tee bolts in pairs with the heads together, turned, screw-cut and faced each end first, then parted in the middle and the heads faced to maximum thickness. They are made from $^7/_8$" b.m.s. and the heads are milled using the indexing attachment. Although I had bought extra change wheels and extra change wheel centres, I found I needed more centres for the change wheels. These are made of b.m.s. and have an integral key $^1/_8$" wide, the diameter being $^5/_8$" for the bore of the change wheels. It seemed a difficult job to make more, but I milled a long length of $^3/_4$" b.m.s. using a $^1/_8$" side-and-face cutter set off from the centre $^1/_{16}$" to the edge of the cutter and $^1/_{16}$" deep and milled around to within a little more than $^1/_8$" of the first cut, then set to the other side and milled back again around the other way, taking one turn of the indexing attachment for each cut. It was then only necessary to bore and ream $^7/_{16}$" and part off to the correct lengths and I had quite a lot of change wheel centres. The bushes, bolts and thick washer were a simple job and I made a number of slotted collars for when one only change wheel is required.

It is possible, and indeed simple, to make what is known as a Cone drive gearing using this attachment. This Cone drive worm gearing is a double enveloping type of worm gear where the worm diameter follows the contour of the worm-wheel. It can be made by fitting a toothed cutter in place of the gear wheel blank to the mandrel and hobbing straight into a shaft which will form the curved worm. The top edge of the cutter is at the centre line of the lathe and the angle of the teeth is at the helix angle of the worm, very similar to the worm-wheel in this fixture.

I think it should be possible to machine a Cone worm using a single point cutter, but I would have to try this out to make sure, which opens up rather exciting possibilities for worm drives. It would be possible to machine a straight worm moving the saddle and locking at a series of pitch centre distances and feeding into the predetermined stop each time. In any case, hobbing of worm-wheels must always be done with the saddle locked, a reading taken as soon as the cross-slide feeds in the blank to touch the hob or cutter and the depth of cut read off the cross-slide feed screw, which is very simple.

With this fixture, the indexing of worm-wheels for such things as milling heads, circular tables and worm feeds for drilling and vertical milling machines can be made with confidence, knowing that the indexing will be correct and accurate.

CHAPTER 17

An Improved Topslide

I have never been happy with the design of the topslide of my Myford Super 7 lathe for a number of reasons:

1. The topslide must be set at an angle in order to get the tool close to the centre without excessive overhang of the tool.
2. When screwcutting coarse threads such as worms or hobs where the slide must be set parallel to the bed so that accurate thicknesses of teeth can be cut, it is necessary to set the tool with excessive overhang.
3. The tool is not well supported due to the round base design even though the tool is set with minimum overhang.
4. The method of locking the topslide by means of two square head set screws is not very secure. Tightening the screws results in altering the gib adjustment of the cross-slide.
5. Swivelling the slide close to 90 deg. results in fouling of the handles of the top and cross-slides so that neither can be used in this position.
6. With the topslide set over to clear the tailstock when turning between centres, the slide cannot be used for its purpose as the graduations are meaningless and any movement at all destroys the readings of the cross-slide.

I finally got thoroughly browned off with these faults and decided to do something about it. I thought at first of a square fixed base, but soon realised that for screwcutting and shouldering to accurate thickness a topslide movement is

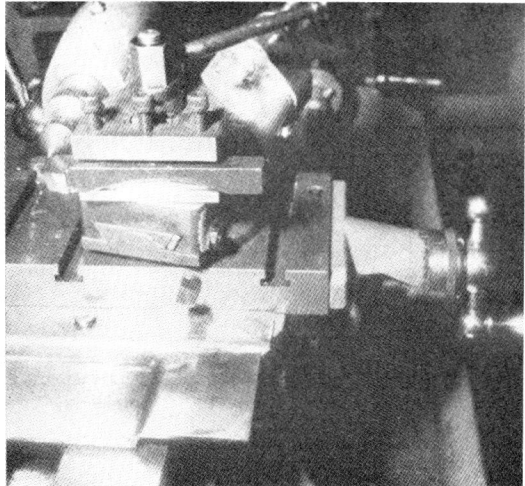

The tool is not too well supported on the existing topslide.

The tool is well supported on the new topslide.

necessary. I realised however that only a very small percentage of turning requires a swivelling slide and that most or nearly all turning could with advantage be done with a fixed topslide where the setting for parallel turning is accurate and automatic.

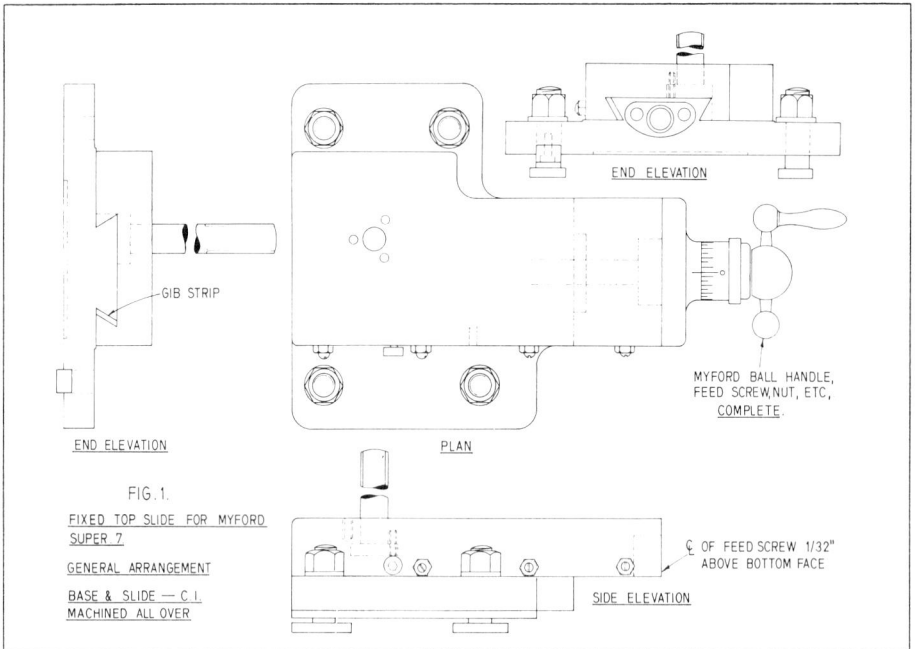

GIB STRIP

END ELEVATION

END ELEVATION

PLAN

FIG. 1.

FIXED TOP SLIDE FOR MYFORD
SUPER 7.

GENERAL ARRANGEMENT

BASE & SLIDE — C.I.
MACHINED ALL OVER

MYFORD BALL HANDLE,
FEED SCREW, NUT, ETC,
COMPLETE.

℄ OF FEED SCREW 1/32"
ABOVE BOTTOM FACE

SIDE ELEVATION

FIG.2.   UNDERSIDE VIEW OF BASE. C.I.   PLAN VIEW OF BASE – MACHINED ALL OVER

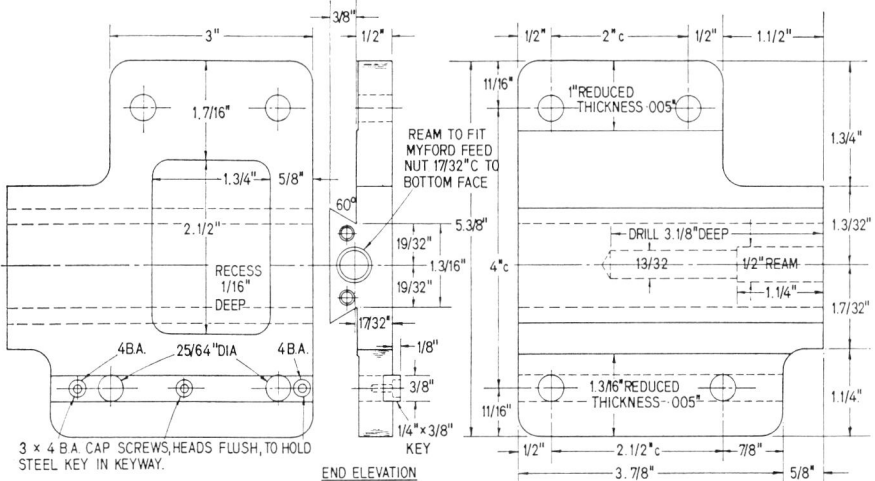

3"   3/8"   1/2"   1/2"   2"c   1/2"   1.1/2"

11/16"   1"REDUCED THICKNESS ·005"

1.7/16"   1.3/4"

1.3/4"   5/8"   REAM TO FIT MYFORD FEED NUT 17/32"C TO BOTTOM FACE

1.3/32"

60°   5.3/8"   DRILL 3.1/8"DEEP

2.1/2"   19/32"   13/32   1/2"REAM   1.3/32"

1.3/16"   1.1/4"

4"c   19/32"   1.7/32"

RECESS 1/16" DEEP

17/32"   1/8"   1.3/16"REDUCED THICKNESS ·005"

4B.A.   25/64"DIA   4B.A.   3/8"   11/16"   1.1/4"

1/4"×3/8" KEY   1/2"   2.1/2"c   7/8"

3 × 4 B.A. CAP SCREWS, HEADS FLUSH, TO HOLD STEEL KEY IN KEYWAY.   3.7/8"   5/8"

END ELEVATION

TEE BOLTS. 4 OFF M.S.

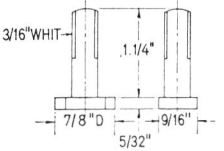

3/16"WHIT   1.1/4"

3/8"   TEE NUTS. 4 OFF. ·525 A/F HEX (5/16"WHIT SIZE)

7/8"D   9/16"   5/8"   WASHERS 4 OFF M.S

5/32"   3/32"   SIDE ELEVATION OF BASE

TOP SLIDE

FIG. 3.   2.5/8"   3.1/4"   9/16"

15/16"   1.5/16"   3/8"

1.1/4"   13/32"   1.5/16"   3/4"   3.1/16"

2.3/4"   3 TAPPED HOLES FOR SQUARE TURRET, 2 B.A.   1.3/32"   19/32"

1.3/16"   2.5/16"   19/32"

5/8"   3/32" STEEL GIB STRIP   1/8"PIN DOWEL   17/32"

5/8"   1"   7/16"   13/16"   13/16"   1.7/16"   3/4"   3/16"

5.7/8"   SPOT THROUGH FROM MYFORD END PLATE

1.1/2"   4 × 3 B.A. GIB SCREWS & NUTS
7/16"B.S.F.   1 × 2 B.A. CLAMP SCREW   3.1/8"   1 × 1/8" DOWEL PIN

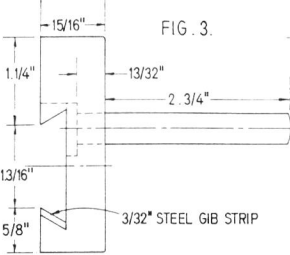

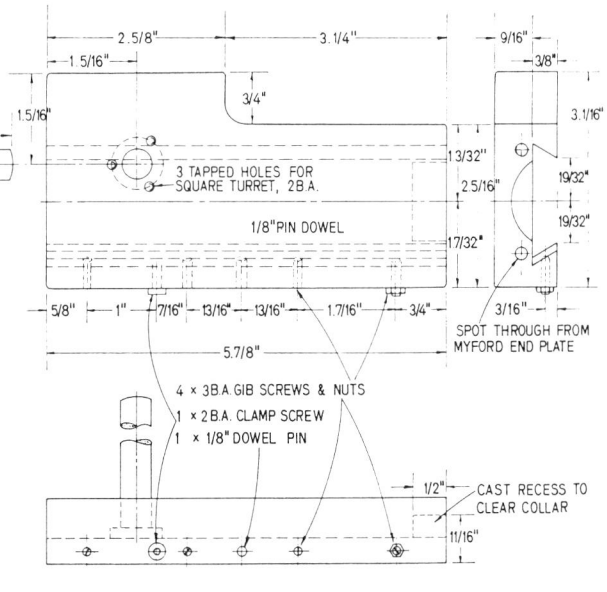

TOOL POST BOLT H.T. STEEL   1/2"   CAST RECESS TO CLEAR COLLAR

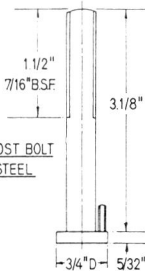

11/16"

3/4"D   5/32"

165

AN IMPROVED TOPSLIDE

The new topslide fits neatly around the tailstock.

It takes but a moment to slide the new topslide off the cross-slide and drop in the original but modified topslide for those short acute tapers where the taper turning attachment cannot be used. For long tapers I use the new topslide with the taper turning attachment by using the extra topslide as

shown in the 'Snippets' chapter at the rear of this book: this makes it very easy to position the tool. I would like to say that of all the fixtures I have made for the Myford lathe, none has given me more satisfaction to make and use than this new topslide. The improvement in finish, accuracy and weight

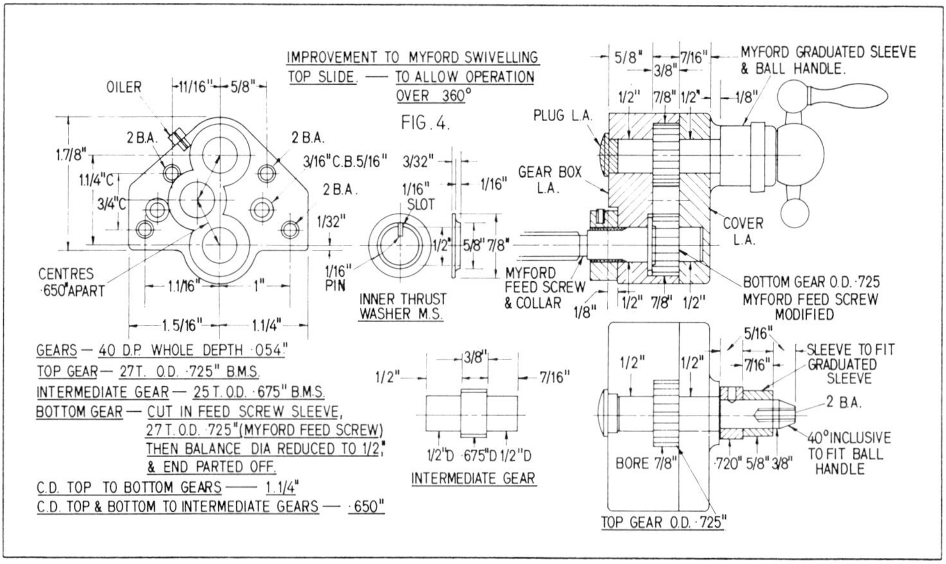

IMPROVEMENT TO MYFORD SWIVELLING
TOP SLIDE. — TO ALLOW OPERATION
OVER 360°

FIG.4.

OILER -11/16"- -5/8"

2 B.A.
1.7/8"
1.1/4"C
3/4"C

2 B.A.
3/16" C.B.5/16" 3/32"
2 B.A.
1/16"
SLOT
1/32"

CENTRES
·650"APART   -1.1/16"- -1"-
-1.5/16"-   -1.1/4"-

GEARS — 40 D.P. WHOLE DEPTH ·054"
TOP GEAR — 27T. O.D. ·725" B.M.S.
INTERMEDIATE GEAR — 25 T. O.D. ·675" B.M.S.
BOTTOM GEAR — CUT IN FEED SCREW SLEEVE,
27 T. O.D. ·725"(MYFORD FEED SCREW)
THEN BALANCE DIA REDUCED TO 1/2",
& END PARTED OFF.
C.D. TOP TO BOTTOM GEARS ——— 1.1/4"
C.D. TOP & BOTTOM TO INTERMEDIATE GEARS — ·650"

1/16"
PIN
INNER THRUST
WASHER M.S.

1/2" 5/8" 7/8"

PLUG L.A.

GEAR BOX
L.A.

MYFORD
FEED SCREW
& COLLAR   1/8"

5/8" 7/16"
3/8"
1/2" 7/8" 1/2" -1/8"

MYFORD GRADUATED SLEEVE
& BALL HANDLE.

COVER
L.A.

BOTTOM GEAR O.D. ·725"
1/2" 7/8" 1/2"   MYFORD FEED SCREW
MODIFIED

-3/8"-
1/2" -7/16"-
1/2"D ·675"D 1/2"D
INTERMEDIATE GEAR

1/2" 1/2"
BORE 7/8"

5/16"
-7/16"-
·720" 5/8" 3/8"
TOP GEAR O.D. ·725"

SLEEVE TO FIT
GRADUATED
SLEEVE
2 B.A.

40°INCLUSIVE
TO FIT BALL
HANDLE

of cut that is possible has to be seen to be believed.

With only $1/2$" overhang of my $1/2$" square tools I can get the tool well past the centre line with the tailstock in position and the overhang of the tailstock barrel is at a minimum, especially as I use a standard tungsten carbide tipped centre which is a little longer than the Myford one. The topslide can be set $1/2$" towards the tailstock if necessary without preventing the tailstock from coming against the saddle. There is ample topslide movement and if turning is done with the topslide, the work is always dead parallel without setting.

The two patterns are very simple; allow a full $1/16$" all over for machining with a little extra on the top of the slide pattern. I machined the top of the slide flat all over as I have an idea for a double clamp for a long 1" square tool holder for maximum diameter turning and backfacing such things as flywheels. The sizes are such that all machining can be done on the lathe, but whatever means are available for milling it is better to face the underside of the base casting in the four-jaw chuck, to prevent distortion.

When renovating my Myford saddle, it left me with 0.001" in 4 inches convex. I was not happy with this as it would gradually get more convex as wear takes place, so I set the saddle 0.0005" **concave** by means of bits of shim packings and machined again to this setting, this time using the lathe bed itself to clamp the shims while the Araldite set. I got a 100% marking immediately this time and guess what, 0.0005" concave in 4 inches! I mention this because, after facing the underside of the base, it is necessary to scrape to a surface plate so that no distortion takes place when clamped for the other milling. The base is $1/2$" thick, but this is reduced a further 0.005" to the edges of where the topslide will fit, so that minimum scraping of the slides is necessary. The slide thickness is $3/8$".

The topslide itself should be left about $1/16$" thicker than the finished $15/16$" so that,

after assembly on the base, the top surface can be machined to exactly the same total thickness of slide and base as the original swivelling topslide and base, so that all tools etc., particularly ball turning graduating tools, will be at exactly the correct centre height.

The slides are machined to 60 deg. and I used the same cutters that I had made for machining the Elevating Heads. The gib piece is $3/32$" thick of ground gauge plate and the gib screws I made 3BA as I considered No. 2 too big and No. 4 too small. A No. 2 cap screw with the head thinned down a little is used as a clamp screw, should it be necessary, and of course a $1/8$" dowel pin is fitted first before drilling for the other screws. The points of the gib screws enter slightly into the gib piece.

When machining the base on the top face, machine the edges at the same time. I used a $5/8$" end mill for the edges which left a nice radius and then the underside can be milled for the $3/8$" × $1/4$" key by setting the machined edge to the dial gauge which will bring the slide dead square so that turning will be parallel. The underside of the slide itself is hand scraped to a surface plate and then the top faces of the base are scraped to this surface. I scraped all the surfaces to fit and then scraped out the marking finally on the raised centre part of the slide of the base. It works so very smoothly that it is a real pleasure to use. All of the machining was completed before scraping however, so that no distortion could take place. I used the feed screw, nut, ball handle etc. from the original topslide and changed them over for use on the original slide until I received a new set of parts for the new topslide. Note that the $1/2$" reamed hole for the nut is $1/32$" above the top face of the base and the hole for the feed screw itself I drilled $13/32$" to $31/8$" depth; the reamed part I made $11/8$" deep. The end plate is spotted through for drilling and tapping 2BA with the slides assembled and lead screw in place.

The central $7/16$" BSF screw for clamping the tool holders is made of H.T. steel, screw

The tool comes well past the centre with plenty of space around the tailstock.

cut and is made with a $^3/_4$" dia. × $^5/_{32}$" thick head and is a tight press fit in the reamed hole in the slide, the balance of the diameter being made a close but sliding fit on the tool holder and square turret. The $^3/_4$" counterbore in the underneath face of the topslide was cut with a home-made spot facing tool before finally reaming the hole to size. After pressing tightly in place, the bolt was secured with a sleeve and nut and then drilled and tapped for a 4BA Allen grub screw through the flange to keep it securely in place, making sure that the screw did not foul the three 2BA screw holes for the 4-post turret. To get the turret dead square, I made a $^7/_{16}$" bolt with a $^1/_{16}$" shoulder in the middle, the shoulder pulling down on the ratchet of the turret with a nut underneath and $^7/_{16}$" dia. above the shoulder, so that the turret could be lined up squarely, the nut then tightened, the turret removed and the three 2BA tapped holes spotted through the ratchet.

The tee bolt nuts I made from 0.525" A.F. hex. steel ($^5/_{16}$" Whit. size) and $^3/_8$" high, tapped $^3/_8$" Whit. The tee bolts were made from $^7/_8$" dia. steel, with shanks $1^1/_4$" long, the heads $^5/_{32}$" thick, milled to $^9/_{16}$" wide. The

holes in the base for the bolts were drilled $^{25}/_{64}$". The key underneath the base is fitted and secured with three 4BA headed Allen screws before drilling for the tee bolts, as the drilling will cut the key into three pieces. The heads of the Allen screws are brought flush with the surface of the key.

One point I should perhaps stress in milling the 60 degree slides is that the tips of the cutter teeth should be ground off a little more than $^1/_{32}$" wide and parallel to the shank of the cutter. Mine are actually $^3/_{64}$" wide, and I have a narrow piece of gauge plate handy for getting the correct width of milling. As soon as the piece of gauge plate starts to enter, do not mill any wider as the scraping of the faces will widen the gap. If the slides must be milled in the lathe, it is better to bolt the castings to the faceplate and use the indexing attachment with the milling attachment on the cross-slide, as the usual vertical slide is a little small for fastening the castings to. All the castings will clear the lathe bed for facing however.

When you finally have everything assembled and mounted on the cross-slide, you will notice how the tool is supported by solid metal right through to the lathe bed and that you have plenty of clearance around the tailstock with the very minimum of overhang of the tool.

Modifying the original topslide so that a full 360 deg. can be used with the two ball handles kept well clear of each other is well worth while for those occasional short tapers and convex and concave turning can then be done. I secured an offcut of 3" dia. Dural (light alloy) rather than bother with castings, as sand castings in light alloy are generally rather porous. The top gear to which is fitted the ball handle is 27 T. and the bottom gear, the lead screw, is milled in the diameter which forms the shoulder inside the graduated sleeve of the Myford feed screw, which happens to be the right size. It is also of 27 teeth, while the intermediate gear is 25 teeth. They are 40 D.P. and the depth for milling is 0.054" 27 T. has an outside diameter of 0.725" and the 25 T. an outside diameter of 0.675". The gears are $3/8$" wide, except that the gear cut in the shoulder is only $5/16$" wide, but this allows for a thrust washer which runs in a shallow recess turned in the gearbox. The gearbox is fully enclosed and is oiled by the

oil gun through a little oiler made from $3/16$" b.m.s., shouldered down to $1/8$", press fitted into the housing with a $1/16$" hole through the oiler. It will be found to be quite oil tight and the b.m.s. shafts run quire happily in the light alloy just as well or perhaps better than they would in bronze.

The gearbox and cover were bored for the gear shafts together in the following manner: The piece of Dural was about 4" long and this was set up in the 4-jaw chuck and faced on the outside, a groove about $3/8$" deep and a full $1/4$" wide was made around the diameter at $1 1/8$" from the face and then was parted off a further $11/16$" along the piece.

The parted-off face was now faced off truly and this face was bolted to the faceplate using $1/4$" thick clamp pieces in the groove made for them. The positions of the holes at $1 1/4$" centres distance for the top and bottom shafts were centre punched, using dividers, the important measurement being the centre distances to the intermediate shaft.

Three discs $3/8$" thick and with a $1/4$" reamed hole in each, one disc 0.625" dia.

Plenty of clearance between the two handles with the gearbox fitted to the original topslide.

and the other two 0.675" dia. (P.C.D. of the gears) were turned accurately and two short pieces of $\frac{1}{4}$" silver steel about 1" long were parted off.

Now with the $1\frac{1}{8}$" wide part outermost from the faceplate and with the lower centre punch mark running truly, the piece was again faced off lightly to make sure and then drilled $\frac{7}{32}$", bored with a small boring tool and reamed $\frac{1}{4}$" to fit the silver steel. The depth of the drilling can be right through in this case as, although we are making the bottom shaft position in the gearbox, it becomes the top shaft position for the cover. Now move the piece over on the faceplate to line up for the top shaft in the gearbox and drill and ream the same, but be careful not to go further than $\frac{1}{16}$" from the face of the cover.

The two 0.675" dia. discs with their two spindles can now be placed in these two reamed holes and the other disc 0.625" dia. with the $\frac{1}{4}$" reamed hole can be clamped to the surface, touching both of the other two discs, which will bring the centre to the correct position. This can now be set to run true and drilled and bored, being careful again not to drill too deep. We now have three $\frac{1}{4}$" dia. holes at the correct centre distances for the gears. The intermediate hole is now in the correct position for boring and can be bored and reamed $\frac{1}{2}$", being careful that it does not go more than $\frac{1}{16}$" from the bottom finished face. This will be slightly more than $\frac{3}{8}$" from the face where the clamps are against, to allow a little for refacing this surface.

The hole is now counterbored $\frac{13}{16}$" to a depth of $\frac{3}{8}$" full to clear the intermediate gear. Now set up to one of the other two holes and get it running true to the dial gauge, using one of the discs if necessary, and bore it out and ream $\frac{1}{2}$" to the same depth as the intermediate gear hole, but counterbore to $\frac{7}{8}$" dia. and $\frac{3}{8}$" depth as before.

The final hole for the feed screw which also forms the hole for the ball handle shaft in the cover can be reamed $\frac{1}{2}$" right through

and counterbored $\frac{7}{8}$", but this time to $\frac{1}{2}$" depth to allow for the thrust washer. Put the piece back in the 4-jaw chuck and part off in the centre of the $\frac{1}{4}$" wide groove. Fit the three-jaw chuck and face off the gearbox half to 1" thick, making sure the machined face is running true against the chuck jaws. A stub mandrel is now made for the cover to fit the $\frac{1}{2}$" reamed hole and the inner face is carefully machined off truly; it can of course be roughed off in the 3-jaw chuck or perhaps it can be finished off here as the other face was true against the faceplate. However, the outer face of the cover must be turned on the stub mandrel to $\frac{7}{16}$" thick, but leaving a $\frac{7}{8}$" dia. shoulder a full $\frac{1}{8}$" thick.

The gearbox is now fitted over the stub mandrel with the inner face to the chuck and counterbored $1\frac{1}{16}$" to $\frac{1}{8}$" deep for the Myford thrust collar.

Two pieces of $\frac{1}{2}$" silver steel can now be used through the box and cover, making sure that the hole for the intermediate shaft lines up on the same side. Before trimming the piece to the final shape outside, it would be better to fit it to the slide for marking around. This is easily done by using the original Myford end cover and fitting it against the gearbox, using the piece of $\frac{1}{2}$" silver steel through the bottom shaft hole and the Myford cover. Spot through for the two $1\frac{1}{4}$" × 2BA Allen cap screws which hold the gearbox to the topslide and drill through $\frac{3}{16}$" and counterbore to fit the Allen screw heads just slightly below flush.

Now mount the gearbox on the topslide and mark around the two sides and the bottom face. The metal left around the top counterbored hole should be $\frac{1}{8}$" and also around the bottom counterbored hole, but with a little flat at the extreme bottom to clear the top of the cross-slide.

The sides etc. can now be machined off by milling, turning or filing; some filing will be necessary to blend the top and bottom radii to the straight sides. The positions for the four $1\frac{1}{4}$" × 2BA Allen cap screws can now be marked out on the face of the gearbox

The gearbox with upper and intermediate gears.

and drilled No. 24 for tapping, but mount the cover with the two pieces of $^1/_2$" silver steel in place and drill through the cover; the cover holes are then enlarged to $^3/_{16}$" and counterbored on the outside to bring the cap screws nearly flush and the gearbox tapped 2BA. Then mounting the cover on the gearbox with the four Allen screws, the cover is finished off to the same contour as the gearbox. I find there is no need for dowel pins, provided these four screws are a nice fit.

The Gears

The top and the intermediate gears with their integral shafts are made of b.m.s. and I made them with an extension of about $1^1/_4$" on one end to hold in the chuck for milling, so that the cutter would clear the chuck jaws. The bottom shaft is the Myford feed screw with the gear teeth milled on the journal which forms the shoulder for the graduated sleeve. A split sleeve reamed $^3/_8$" dia. was used to hold it in the chuck and after milling the $^5/_8$" journal next to the gear teeth, it was turned down to $^1/_2$" dia. by $^3/_8$"

long to fit the reamed hole in the cover. A $^1/_{16}$" pin was fitted just immediately back of the gear teeth which forms a drive pin for the inner thrust washer which is made with a $^1/_{16}$" keyway and is cut away on the gear side to clear the intermediate shaft gear teeth. The end of the feed screw shaft is parted off flush with the $^1/_2$" dia. journal.

A sleeve with integral thrust collar, a replica of the end of the Myford feed screw, is made with a $^3/_8$" reamed hole to fit the $^3/_8$" part of the top shaft and this is secured with a 2BA Allen grub screw. This part is to form a journal for the graduated sleeve and the end is made 40 deg. inclusive to fit the ball handle. The end is tapped to fit the ball handle screw 2BA.

I used the new topslide for making all these parts, including the tee bolts, using temporary tee bolts and this top gear shaft was one of those rare jobs which require a swivelling topslide. So this was a good chance to see how long it takes to slide off the new topslide and fit the old one for the short taper only. It took just 40 seconds.

CHAPTER 18

A Thread Milling Attachment

Screwcutting in the centre lathe with a single point tool is passing into the limbo of forgotten things in commercial practice, along with chipping and filing, milling of gears in the milling machine and hand scraping.

Such things will, however, remain with the home workshop enthusiast for a very long time. In the case of screwcutting for such things as feed screws, especially of coarse pitch and perhaps multi-start threads and the making of worms in D.P. pitches and threads etc. in metric pitches, where the lathe must be reversed to return the saddle, the work can become very tedious. This is especially so on deep or coarse threads, where accuracy and finish must be good. One has to take very light cuts towards the last, for fear of a dig-in or a torn thread.

Commercially rolled threads even of coarse pitch are now common practice, milled threads and threads ground from the solid have all replaced the old screw-cutting methods. Rolled threads and threads ground from the solid are I think out of the question for the amateur's equipment, however it is quite possible to mill threads of any pitch in multi-start or long lead in the lathe with one pass per thread and to excellent accuracy and finish.

I have designed and made such an attachment that will mill a thread of any helix angle up to 30 deg. each side for left and right-hand threads and with a 2" dia. cutter will mill down to zero with the tailstock in place and with minimum extension of the tailstock barrel. It will also mill with the travelling steady in place; in fact the cutter

The completed thread milling attachment.

comes to the centre of the travelling steady jaws, thus there is no twisting or other strain on the work, allowing very slender work to be milled. The cutter remains on the centre line of the lathe when moved to any angle and so the thread remains accurate.

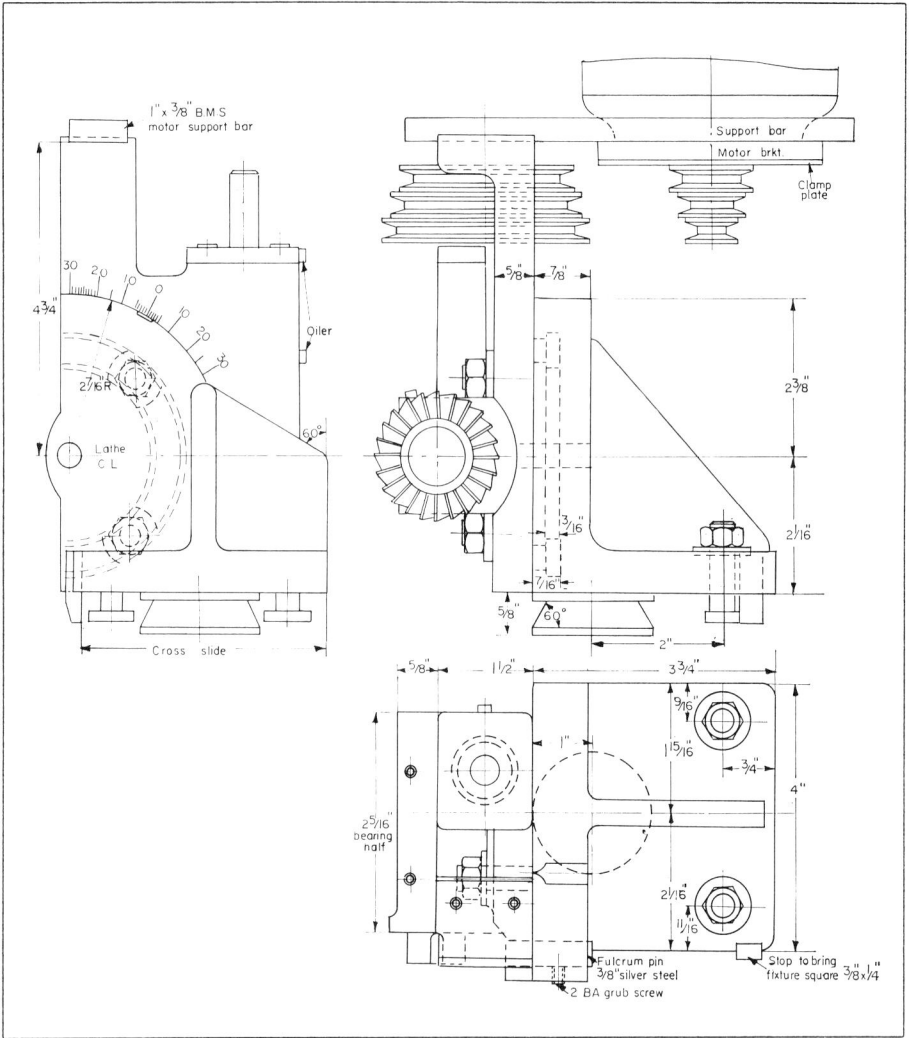

Milling of a thread right up to a shoulder is easily accomplished and can be taken very close to the chuck jaws or carrier. The cutters used are standard 2" dia. thread milling cutters of $^7/_8$" bore and, while I make my own, they should not be difficult to obtain. Cutters of fine tooth type should be used. I make mine of 24 teeth, and up to $^1/_2$" wide can be accommodated. The cutters are of single row type. The direction of the cutter is upwards at the cutting point, thus all strain is taken in the normal downwards direction.

In designing this fixture, a lot of thought

Milling a ¹/₂" pitch, four start thread.

was given to bringing the cross-slide to the normal turning position without excessive overhang from the saddle. This made it necessary to use worm gearing for the drive, although I would have preferred a gear reduction such as my milling fixture (Chapter 1). I have used the same motor, motor bracket and belt from this fixture and I was very agreeably surprised, on testing it, to find that I had ample power left for milling and that the finish was excellent with no worries of dig-ins or torn threads.

Four simple patterns are required, the main bracket of heavy construction to allow for the semicircular tee slot in the vertical part and with the bottom circular boss to fit the Myford cross-slide. The swivelling main casting has a ⁷/₈" core for the worm shaft, but is left flat on the front face for the cutter shaft, the cover for the cutter shaft and the pulley – a replica of the pulley on the milling attachment but bored and reamed ⁷/₁₆".

The main bracket is set up on the faceplate and bolted to an angle bracket for turning the bottom face and the circular boss which is a replica of the topslide; the

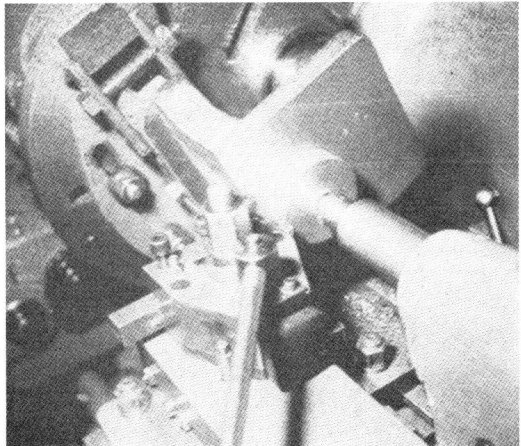

Machining the bottom face of the angle bracket.

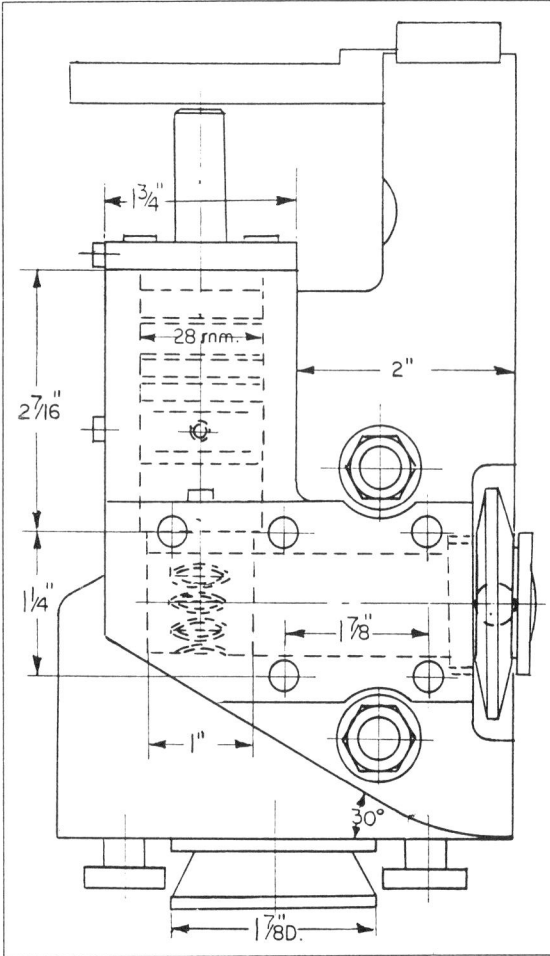

two faces are hand scraped, the bottom face to the cross-slide and the vertical face to the surface plate. It is now necessary to drill for the fulcrum pin exactly on the centre-line of the lathe, so set up the bracket on the cross-slide with the machined face to the tailstock and bring square to the cross-slide; put a drill chuck in the lathe headstock and a large centre drill in the chuck and drill in the position as shown, follow up with a $^{23}/_{64}$" drill and then ream $^3/_8$" dia. for the silver steel fulcrum pin. At this same setting and using a large diameter fly cutter, true up the edge of the base casting, as this will assist in resetting the casting on the faceplate for milling the semi-circular tee slot.

Now fit the faceplate on the lathe and with the fulcrum pin in the tailstock chuck, reset the casting on the faceplate, bolting to the angle-plate and supporting it in position on the fulcrum pin. It is only necessary to get a complete half turn over the lathe bed for this milling operation, so there will be no trouble in clearing the lathe bed. Check for true running on the fulcrum pin when secure with the dial gauge. I had forgotten to say that it will be found better to drill for the two tee bolts in the position shown, as these holes will be found useful

edge of the boss should be no more than $^1/_8$" from what will be the finished vertical face. When set up correctly, it will clear the lathe bed easily; centre with a large centre drill before turning and no trouble will occur in getting accuracy and good finish, using the tailstock for support.

Face the vertical face now in the centre of the faceplate, using the angle-plate again in a position to clear the lathe bed. These

for bolting up. Make sure the casting is square before drilling these holes by bringing the vertical machined face against a true mandrel or bar in the lathe, and marking underneath with a bent scriber.

The milling of the semi-circular tee slot is done with the milling attachment on the fixed vertical slide and the $^3/_8$" end mill set to $2^3/_{16}$" radius above the fulcrum pin. Mill down to $^7/_{16}$" depth using the indexing

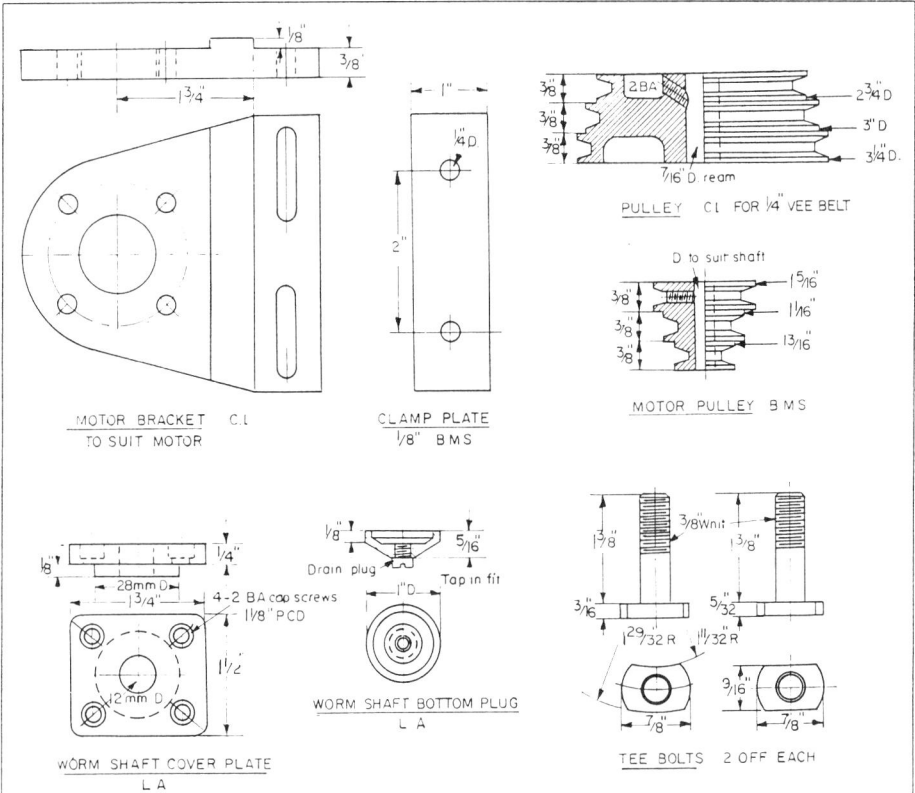

MOTOR BRACKET C.I
TO SUIT MOTOR

CLAMP PLATE
1/8" BMS

PULLEY CI FOR 1/4" VEE BELT

MOTOR PULLEY BMS

WORM SHAFT COVER PLATE
L A

WORM SHAFT BOTTOM PLUG
L A

Drain plug

Tap in fit

4-2 BA cap screws
1 1/8" PCD

TEE BOLTS 2 OFF EACH

attachment (Chapter 2) as a driving mechanism. Then replace the end mill with a $^3/_{16}$" × $^9/_{16}$" dia. tee slot cutter and follow through with this to just clear the bottom of the groove; this should leave $^1/_4$" from the face to the inside edge of the tee slot. To complete the main bracket, the stop to bring the bracket square with the lathe centres should be made and fitted from $^3/_8$" × $^1/_4$" key steel secured in a shallow groove with a 4BA Allen grub screw. Using a true mandrel in the lathe centres, the tee bolt holes should be spot faced and, if necessary, the tee bolts can now be made, as these will be required for subsequent operations. After I had milled the tee slot, I also ran the end mill lightly over the top curved face to make

sure that this face conformed to the proper shape.

The swivelling main casting should be machined to a total thickness of $1^1/_2$" and both faces scraped to the surface plate. The bottom end at 30 deg. should now be faced and then the fulcrum pin hole can be drilled and reamed. This is drilled so that the centre will overhang the cross-slide by $^1/_4$" so that the cutter, which is always spaced centrally on the spindle, will come to the centre of the travelling steady jaws.

With the fulcrum pin in place, the two tee bolt holes at $2^5/_{16}$" centres can be drilled, by marking with a bent scriber from the main bracket. These are drilled $^{25}/_{64}$". With a $^3/_4$" end mill, clearance for the tee bolt nuts and

Drilling for the main spindle.

a $^5/_{16}$" Whit. ring spanner is provided and at the same time the two raised bosses for the tee bolt nuts are milled. These clearances need be milled in the main casting only and need not be milled in the bearing cap, as the ring spanner can be slipped over the tops of the nuts. The nuts for these two tee bolts are made from $^5/_{16}$" Whit. hexagon steel and the heads of these two tee bolts are curved to fit the tee slots.

The bearing cap is faced off on both sides to $^5/_8$" thickness except that the end next to the cutter is $^7/_8$" thick as this thicker part will clear the tailstock centre when set up for milling a thread. Before drilling for the

five 2BA cap screws which bolt the bearing cap to the main casting, set up this casting in place on the cross-slide and bolt to the main bracket in a perfectly vertical position as the spindle bore will be bored in this position and it is necessary to get the correct position for the Allen screws so that they will not foul the spindle bore. Bolt the cap into position with a couple of very thin pieces of shim stock, so that subsequent wear can easily be taken up. Now set up the whole job in position on the cross-slide in a perfectly vertical position.

I machined the edge of the swivelling casting to make this easy and, with the joint

Boring for the worm spindle bearings, using the author's elevating heads.

Facing the main spindle hole.

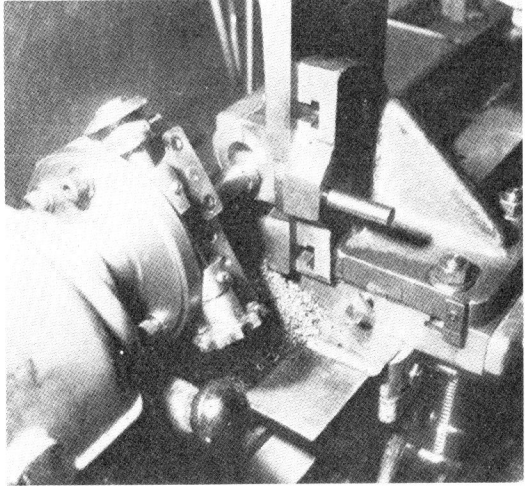

of the two faces on the centre line of the lathe, locked the cross-slide in position. The main bore can now be centre-drilled and drilled with a large drill, I used a $^{25}/_{32}$" taper shank drill and then followed with a boring bar and bored accurately to $^{7}/_{8}$" and to a good finish. The front end is faced off and to do this, I used my automatic facing head (Chapter 15). It is then counterbored $1^{1}/_{4}$" to $^{5}/_{32}$" deep and the opposite end also counterbored to $1^{1}/_{4}$" to a distance of 3" between the two faces. Be careful therefore

when drilling for the five 2BA cap screws, that you allow for these two counterbored faces. While in this position, the bore can also be polished using a piece of fine carborundum strip secured through and around a piece of dowelling.

The worm shaft runs in four No. 6001 ball races and the upwards thrust of the worm is taken by a thrust race No. 51101, so the main casting should now be set up for boring the $^{7}/_{8}$" cored hole to 28mm at exactly 0.750" centre distance from the

Milling the semi-circular tee slot in the angle bracket.

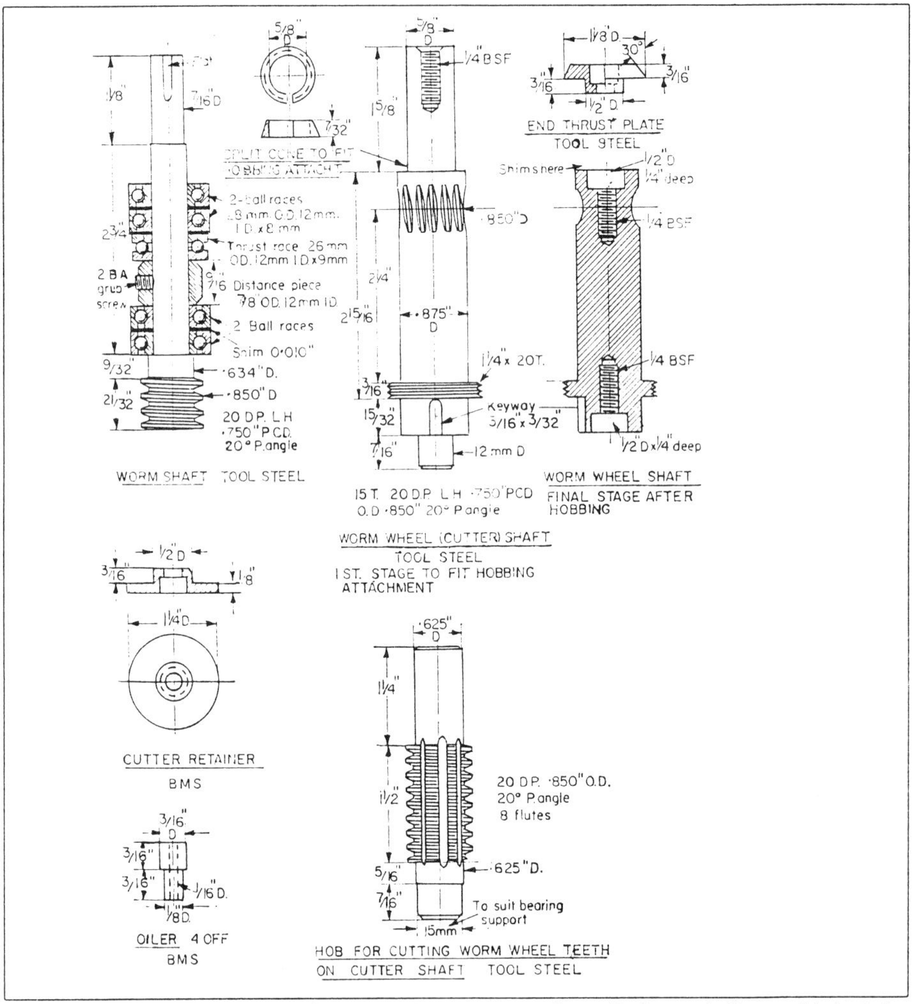

SPLIT CORE TO FIT HOBBING ATTACHMENT

2-Ball races
9 mm O.D. 12 mm
I.D. x 8 mm
Thrust race 26 mm
O.D. 12 mm I.D x 9 mm
Distance piece
78 O.D. 12 mm I.D
2 Ball races
Shim 0·010"

2 B.A. grub screw

·634"D.
·850"D
20 D.P. L.H.
·750" P.C.D.
20° P. angle

WORM SHAFT   TOOL STEEL

END THRUST PLATE
TOOL STEEL

1/4 BSF

·850"D
·875"D
1/4 x 20T.
Keyway
5/16 x 3/32
12 mm D.

15 T. 20 D.P. L.H. ·750'PCD
O.D ·850" 20° P. angle

WORM WHEEL (CUTTER) SHAFT
TOOL STEEL
1 ST. STAGE TO FIT HOBBING
ATTACHMENT

Shims here
1/4 BSF

WORM WHEEL SHAFT
FINAL STAGE AFTER
HOBBING

CUTTER RETAINER
BMS

OILER   4 OFF
BMS

20 D.P. ·850"O.D.
20° P. angle
8 flutes

·625 "D.
To suit bearing
support
15mm

HOB FOR CUTTING WORM WHEEL TEETH
ON CUTTER SHAFT   TOOL STEEL

main shaft bore just finished. This can be set up on the cross-slide, but I found it very simple to use my elevating heads and quartering table (Chapter 7) to get the right centre distance and the two bores square. The main spindle (which I had previously completed) was set between centres, the main casting was laid on the quartering table, the spindle lowered into the bore (cap removed) until the spindle could just be turned, the main casting bolted up to the quartering table, the spindle removed, quartering table turned 90 deg. and the spindles lowered 0.750", and with the centres at 2¼" from the inside of the counterbored face, the boring for the worm shaft could commence.

The boring can also be done directly

Graduating the degree marks of the main casting.

bolted to the cross-slide and with suitable packing, accuracy can be assured. The bore is bored to exactly 28mm at $2^{1}/_{4}$" from the inside counterbored face and down to within $^{5}/_{8}$" of the centre line of the main bore, the balance of the bore being 1". This will leave a shoulder for the bottom ball race. The top face of this bore is faced off for the cover plate at $2^{7}/_{16}$" from this shoulder. One more thing remains to be done to this main casting and that is the graduating of the degree marks for setting the cutter to the helix angles. Set up the casting on the small faceplate directly on the machined bearing

face using the fulcrum pin in the drill chuck in the tailstock as before.

The top lug for the motor support bar will clear the small faceplate but not the large one and, after marking the correct commencing position for these graduations from the bracket casting, make a shallow groove about $^{1}/_{8}$" wide to clear the graduating tool below this line, and commence with the 30 deg. line as a horizontal line dead on the centre line of the fulcrum pin. 60 degrees are marked out, the individual degree lines are $^{1}/_{4}$" long, the 5 deg. lines $^{3}/_{8}$" and the full 10 deg. lines $^{1}/_{2}$" long. I used my graduating

Hobbing the worm wheel using the worm wheel hobbing attachment.

Form relieving the teeth of the worm wheel hob.

tool (Chapter 2) for this job. The top lug for the motor support bar is machined 1" wide × $^1/_8$" deep at $4^3/_4$" from the centre of the fulcrum pin and this is drilled and tapped for a couple of $^1/_4$" Allen Whit. cap screws.

The pulley casting is made with a chucking boss on the larger side and this is machined first, and then the pulley is completely machined all over before this boss is turned off. The motor pulley, of mild steel, is the original one made for the milling attachment, as is also the motor support bracket and support bar, with the clamp plate of 1" × $^1/_8$" b.m.s.

It remains now to make the main shaft (the worm wheel shaft) and the worm shaft and also a hob to hob the main shaft. Make the hob first of oil hardening tool steel, the o.d. is 0.855", the pitch is 20 d.p. left hand, depth 0.108" and with eight milled flutes. The pressure angle is 20 deg. This is used in my Worm Wheel Hobbing Attachment described in Chapter 16 and in order to hob in this attachment, it is necessary to make the main shaft longer at each end to hold in the attachment and also to make a little tapered split collar to fit. These dimensions are clearly shown on the drawing. The o.d. for this worm wheel is 0.850", that is 0.0125" below the surface of the spindle, and the depth is also 0.108", hobbing is left hand and 15 teeth are hobbed. This main spindle

is also made from the same oil hardening tool steel and there is a 20 t.p.i. R.H. thread on the edge of the $1^1/_4$" dia. shoulder which will keep swarf and cutting fluid out of the bearing. After the teeth are successfully hobbed, the two ends are modified, but only after cutting the $^3/_{16}$" × $^3/_{32}$" deep keyway for the cutter. The two ends are drilled and tapped for the $^1/_4$" BSF cap screws and counterbored $^1/_2$" for the retainer covers. The cutter retainer can be made of b.m.s., but the end thrust plate is made of the same tool steel. A few thin shims are placed between the end of the spindle and the thrust plate to take up end wear.

The worm shaft is also made of oil hardening tool steel and the worm is 0.850" o.d. 20 d.p. left hand and depth 0.108" and 20 deg. pressure angle. The worm is $^{21}/_{32}$" long, the shoulder next at $^9/_{32}$" long and 0.634" dia. (the root size of the worm). The length for the bearings is $2^3/_4$" at 12mm dia. and the balance $1^1/_8$" long at $^7/_{16}$" for the pulley.

There are two 0.010" shims 28mm × $^7/_8$" dia. between the top two bearings and the thrust bearing, and one 0.010" shim $^5/_8$" × 12mm between the bottom two bearings. A distance collar $^7/_8$" o.d. × 12mm i.d. and $^9/_{16}$" long goes below the thrust bearing and above the second bearing as shown on the drawing. The thrust bearing must go the

The author in his study. In the background can be seen two of his prize-winning models, a beam engine and a 3½" gauge 4-6-0 'Black Five' locomotive.

right way up, the largest outside diameter goes to the top and the plate with the largest hole goes to the top. The distance collar is provided with a 2BA grub screw let slightly into the shaft which will prevent the worm shaft from working downwards at any time. A flat for the pulley is provided. Both these shafts are hardened outright in oil and are then drawn to a medium straw with the worm shaft, which can be left soft on the shank by not heating to full redness when hardening.

The cover plate of light alloy is fitted with four 2BA Allen cap screws at $3/4$" radius at 90 deg. The bottom shoulder is 28mm and is $1/8$" deep. This depth can be adjusted when machining to take up all end play without making the bearings in any way tight when screwed down. A bottom cover plate is

made a light tap fit in the 1" dia. end of the worm shaft hole and is provided with a drain plug. It can be firmly secured with a little Araldite.

Four little oilers just about complete the attachment, these are made as previously from $3/16$" b.m.s. rod, shouldered down to $1/8$" and drilled $1/16$", one to be fitted into the cover plate, one in the worm shaft bore opposite the distance piece and two for the main spindle. In use it will be found quite possible to keep the worm drive flooded with oil and, with both being of hardened tool steel, the drive should wear very well indeed. The hardened main spindle running in cast iron should also last a long time and it is also very rigid.

To drive the lathe spindle and thus the job, even the slowest speed in back gear is much too fast as this drive is virtually a milling feed motion, and so I use the indexing attachment as a hand feed drive. It would be a good suggestion to make a little motor-driven worm drive to the bull wheel of the lathe constructed in a similar manner to the indexing attachment using, say, a sewing machine motor with belt drive to the worm shaft and the foot control as a speed regulator. The drive is very light, the torsional strain on the job is considerably less than with a single point tool and this makes it possible to cut very much coarser threads without strain on the tumbler or change gears.

To get the correct angle at which to set the cutter so that the cutter clears the sides of the thread is easily calculated, an example of a four threads per inch on an outside diameter of $5/8$" in Acme thread is shown. This would be extremely difficult to cut with a single point tool but quite simple with this attachment.

$$\frac{\text{Tangent of}}{\text{lead angle}} = \frac{\text{Lead of screw thread}}{\text{Pitch circumference of thread}}$$

Pitch circumference = O.D. – depth of thread × 3.1416.

4 t.p.i. $5/8$" O.D. Acme thread =

$$\text{Tangent } \frac{0.250 \text{ (lead)}}{0.500 \times 3.1416}$$

(depth of thread is 0.125"),

$$\therefore \text{ P.D.} = 0.500"$$

$$= \text{Tangent } \frac{0.250}{1.571}$$

$$= \text{Tangent } 0.159 = 9 \text{ deg. } 3'.$$

CHAPTER 19

Renovating a Myford Lathe

I received my Myford Super 7B long bed lathe in June 1962 and it has been in constant use since, mainly on work that could be considered a little heavy for this size of lathe perhaps; but it has given me wonderful service. It has been used for work much more varied than straight turning as some of my articles will indicate. It has made me a ¹/₂" drilling machine, a shaper and a power hacksaw and converted a planer to a planer-mill, and all those fixtures and more, that have been published in this book.

Lately however the saddle, when adjusted correctly close to the chuck, was too tight further along the bed and facing has developed a convex tendency. On measuring the amount I found that it faced 0.001" per inch of cross travel convex. Adjusting the saddle to slight stiffness, I found I could still 'rotate' the saddle slightly but visibly.

The inside edge of the front shear of the lathe bed takes the full thrust and wear when external turning and the inside step of the saddle which wears against this edge is only 4³/₄" long, whereas the gib strip at the front of the saddle which takes the wear when boring, is 7¹/₂" long. This design is done no doubt to conform to the modern idea of the narrow guide principle, but with this design of lathe saddle, in my humble opinion, it is not very good.

My drawing will show what takes place under the side thrust of a normal cutting tool when the saddle is drawn along by the leadscrew. The right-hand grub-screw at the front of the saddle adjusts against nothing at all, and most of the wear takes place on the right-hand end of the central step under the saddle and the left-hand end of the gib strip, resulting in convex facing, swivelling of the saddle due to convex wear on this central step and wear on the inside edge of the lathe bed, allowing looseness of the tailstock when close to the chuck.

The amount of wear on the bed was barely measurable so the main wear must have been on the inside step of the saddle. I was now faced with the decision to buy a new lathe, buy a new bed and saddle, or repair the existing lathe. As the rest of the lathe was in excellent condition I decided on the last possibility. With the apron removed and the saddle on the unused right-hand end of the bed, I found that the space between the machined rear edge of the rear shear of the lathe bed and the machined but unused inside rear vertical face inside the saddle measured 0.033" at the left hand end and 0.027" at the right hand end, a difference of 0.006" in 7¹/₂" length. Here then was proof of the convex facing due to wear and that the original machining of the saddle was correct. It seemed a pity to have a perfectly good machined face on the rear edge of the bed that is not used for any purpose at all.

I found the gap in front of the saddle gib strip, where wear had been taken up from time to time, now showed a space of 0.033" at the left-hand end and 0.035" at the right-hand end. All of these measurements were taken at the right-hand end of the bed which was almost unused and would have been accentuated close to the chuck. Removing

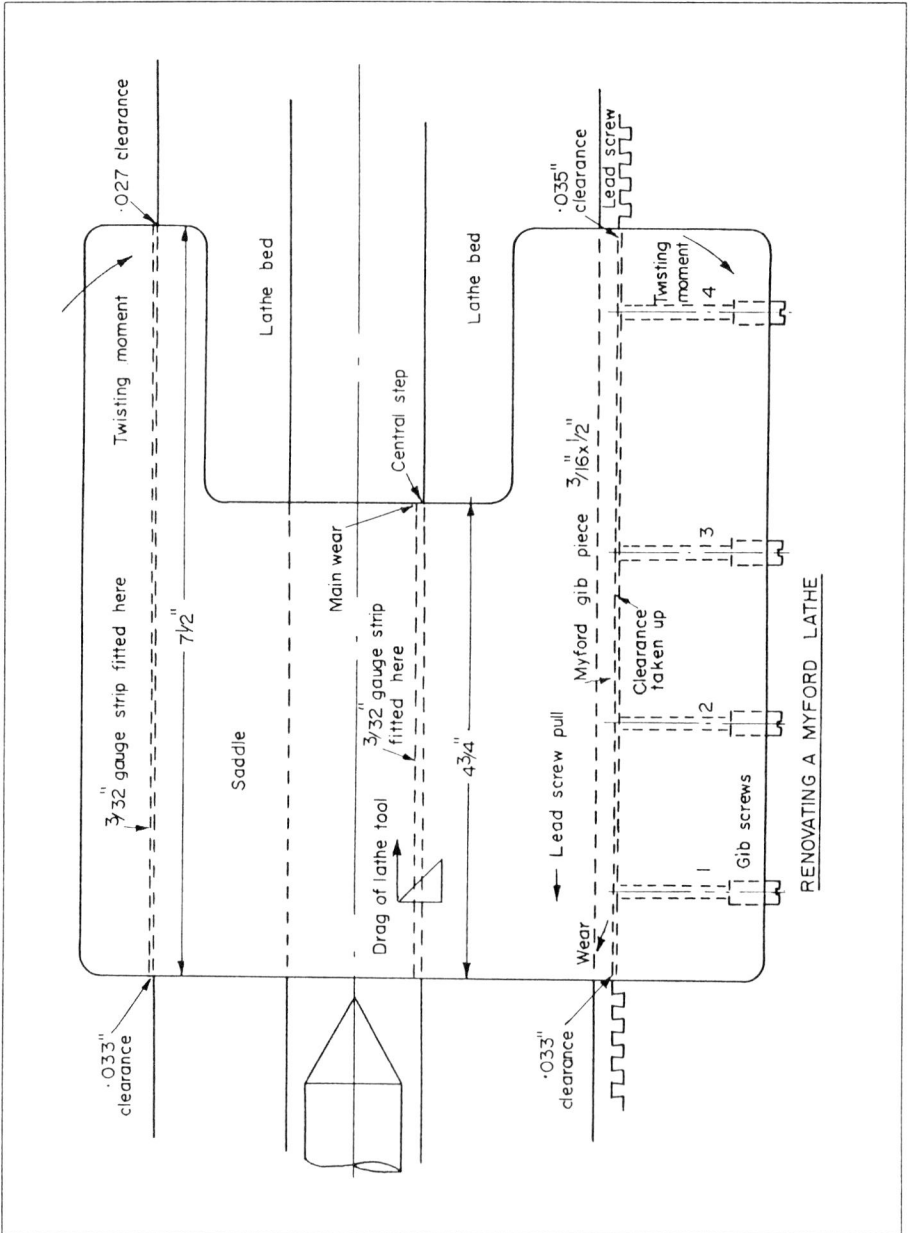

RENOVATING A MYFORD LATHE

the saddle, the curved wear on the central step was very apparent.

I had a piece of $^3/_{32}$" ground gauge plate, so I made one gib piece $7^1/_2$" long by $^1/_2$" wide and one piece $4^3/_4$" long by $^9/_{16}$" wide. Now if I machined the inside face of the rear edge of the saddle, by taking up all the space between the front gib strip and the saddle, I would need to machine out 0.034" to take the gauge strip and leave a couple of thou clearance and remove 0.062" from the centre step face to allow for the other gauge strip. This I did setting up to the dial gauge on the unused inside edge of the rear of the saddle. The work was done on my plano-mill using a good end mill, set to just clear with a feeler gauge of 0.002" between the end and the face of the saddle.

Using marking blue on the rear edge and the rear inside edge of the lathe bed, I checked for marking on both faces. My central step was a little high, so I scraped it down until I had a true face on both edges of the saddle. Next I clamped the central gauge plate in position against the central step, and after making a long No. 32 and $^9/_{64}$" drill, I drilled right through the front gib screw holes and into and through the gauge strip and the central step with the No. 32 drill, removed the gauge strip and drilled through the central step only with the $^9/_{64}$" drill.

I now tapped the gauge plate 4BA and set up the saddle in the machine vice on the drilling machine again, with the rear edge of the saddle to the top and drilled No. 24 through the rear edge of the saddle and the long gauge strip, after clamping with a long piece of $^3/_4$" key steel as before. These holes were drilled right in line with the front gib screw holes.

The long gauge strip was now removed and the holes through the saddle edge only were redrilled $^3/_{16}$" and the long strip tapped 2BA. A drill was prepared in the form of a spigot drill from silver steel that would go through the $^3/_{16}$" holes and the curved face of the central step was counterbored to take the heads of the 4BA Allen screws which hold the gauge strip to the central step.

I now spread a thin layer of Araldite on the central step and the inside face of the rear of the saddle and a thin layer on the gauge plate strips and clamped them both up overnight using the $^3/_4$" key steel with the Allen screws in place but not tightened. In the morning the Alien screws were fitted, three 4BA through the central step and four 2BA through the rear edge of the saddle, and tried again with the marking blue. The marking was not perfect but was showing quite well so with the end of a fine! file, I 'scraped' the surfaces of both strips to very nearly 100 per cent marking on the unused right hand end of the bed.

I then fitted the front gib strip and adjusted carefully, first No. 2 gib screw, then No. 1 and No. 3 simultaneously and lastly No. 4, which now had something to adjust against. I could now slide the saddle beautifully the full length of the bed by hand and there was absolutely no movement anywhere of the saddle close to the chuck or at the extreme right hand end of the bed. I was able to remove one layer of the laminated shim stock from the underneath plates under the saddle, and this fortunately made no difference to the feel of the saddle, although all vertical play was removed.

My saddle was now $^1/_{32}$" further towards the rear, but the holes for the Allen screws were drilled with plenty of clearance for the apron and this was set in the correct position by gripping the leadscrew tightly with the leadscrew nut close to the head-stock, which determined the lateral position of the apron. After tightening the apron, the nut was opened and closed several times, but no movement of the leadscrew was discernible, so I had sufficient clearance in the apron screws.

I now had an area $12^1/_4$" × $^1/_2$" to take the thrust of external turning instead of $4^3/_4$" × $^1/_2$" and the twisting moment of the lathe tool now operated against the rear edge of the lathe bed at a considerable distance from the tool instead of being very close to it, which results in very little pressure and wear and it will be a long time before further wear takes place on the inside edge of the bed, so the tailstock will not be further affected.

I checked the facing with the dial gauge and found it was less than 0.001" convex in 4"; this amount is possibly due to some wear in the cross-slide, but is so small that it does not matter. The rest of the lathe is in perfect condition, so I now am back to new again, or dare I say it, better than new. I am so pleased with the lathe now, indeed, that if I did ever buy a new Myford, I would make these modifications to the saddle before I even started to use the lathe.

CHAPTER 20

Snippets

Readers may be interested in photographs of a Relieving Attachment and a Speed Reduction Head from which the Relieving Attachment is driven. In photograph No. 1 it will be seen that the cross-slide screw and handwheel has been removed and feed to the tool is taken from another slide mounted to work in a mounting block bolted and dowelled to the topslide. The cross-slide reciprocates to the number of flutes under the action of a double cam. The cam is located in a large ball race which is rigidly secured to the saddle in a bracket, the cam running against a pair of hardened rollers secured to the cross-slide in a solid steel plate fitting which can be bolted to any of three positions to suit different diameters of cutters. The ball race is driven by a double keywayed $5/8$" diameter shaft which is driven from a train of gears in a quadrant bolted to the left-hand end of the Speed Reduction Head. Any number of flutes within reason can be relieved. The speed reduction of 1:6 of the head is necessary and, with the lathe in back gear, the reciprocation speed of the cross-slide is quite normal. Also, it makes possible by this speed reduction the forming of hobs for gear and worm wheel cutting of any pitch front 3 d.p. to 120 d.p. where normally the coarsest pitch was 16 d.p. and this without putting any extra strain on the tumbler gears. In this photograph the plain cover of the head has been removed. In the second photograph a geared head is shown complete with a dividing plate, etc., and enables helical gears to be cut from about 2" lead up to more than 300" lead. The speed reduction in this case is 1,080 to 1.

A further attachment for this Speed Reduction Head has been designed and is contemplated where the lathe is turned into

Photograph No. 1

Photograph No. 2

a complete hobbing machine for plain and helical hobbing of gears. I may not however make this, as helical milling as in photograph No. 2 is so completely satisfactory using the hob type milling cutters described by me in Chapter 4, but which are now properly radially relieved as in photograph No. 1.

I trust that some of the more advanced readers will be interested in the above set-ups which are not generally found in the workshops of the amateur.